THE
MINI
STORY

THE
MINI
STORY

herausgegeben von / edited by
Andreas Braun

HIRMER

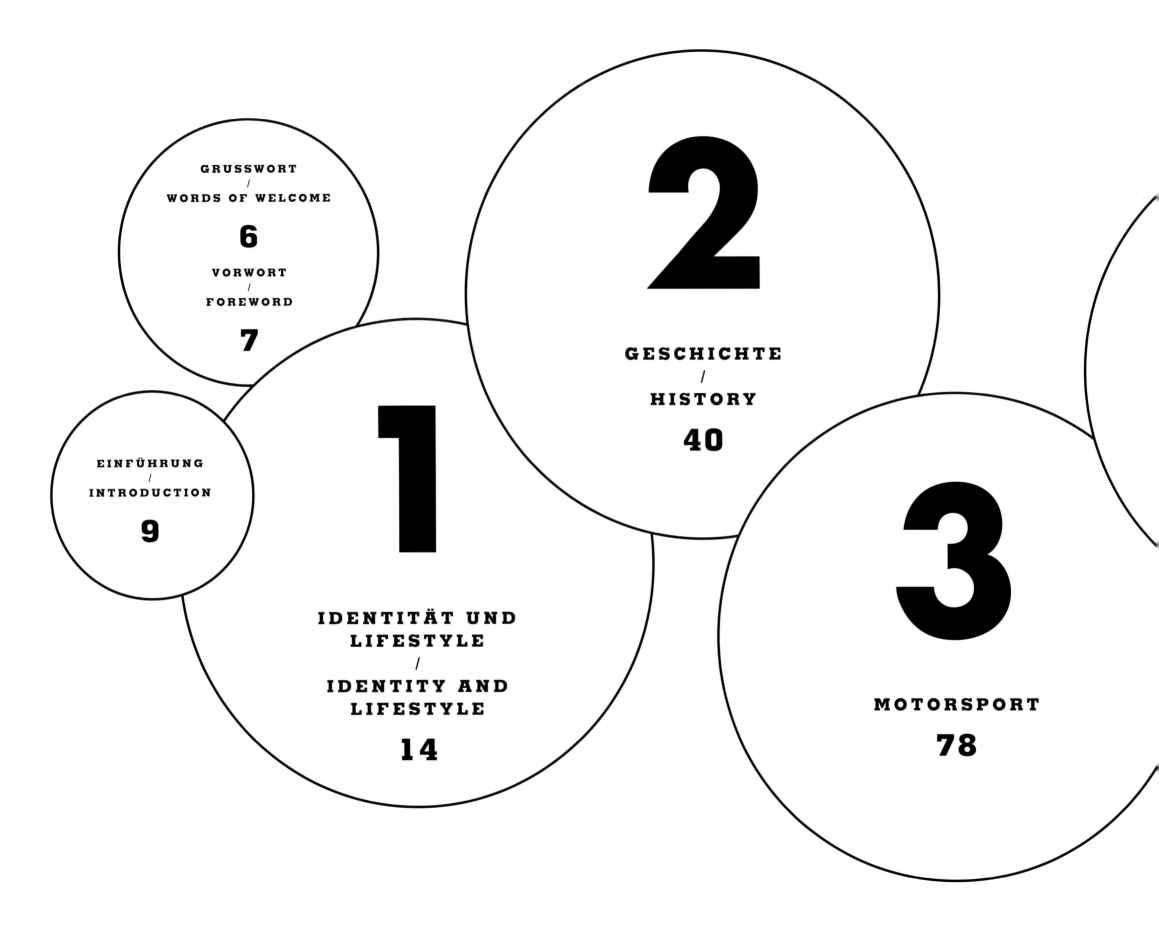

Ein Automobil, das den Namen MINI trägt, ist mehr als die Summe seiner Einzelteile, mehr als ein langlebiges Konsumgut oder ein charmantes Fortbewegungsmittel. Wie es unsere Kunden immer wieder bestätigen, hat der MINI einen ganz besonderen Charakter.

Jeder MINI inspiriert und verbindet Menschen auf der ganzen Welt. Die Marke überwindet Grenzen und setzt Trends. Die Bindung zu einem MINI ist immer vor allem eine emotionale. Der Slogan »like a friend« ist das Bekenntnis von tausenden Fans in aller Welt, die MINI in ihrem Leben nicht missen wollen.

Auch ich habe eine persönliche Bindung zu dieser außergewöhnlichen Marke. Vor vielen Jahren machte ich mit einem liebevoll restaurierten klassischen Clubman die nächtlichen Straßen Madrids unsicher. Diese Erfahrung, das Gefühl, das mir der Wagen vermittelt hat, und die große Sympathie der Menschen für mein Gefährt werde ich nie mehr vergessen. Der Clubman als mein Begleiter hat diese Fahrten zu einem besonderen Erlebnis gemacht. Ein weiteres Bindeglied zwischen mir und MINI ist natürlich auch das gemeinsame Geburtsjahr: 1959.

MINI hat über Jahrzehnte bis heute Automobilgeschichte geschrieben. Tauchen Sie mit diesem Bildband in die Geschichte von MINI ein. Lassen Sie sich von den unterschiedlichen Facetten dieser Marke begeistern und erleben Sie Seite für Seite ihre Entwicklung von den Anfängen bis zur Gegenwart. Ich wünsche Ihnen viel Vergnügen mit den teilweise erstmals veröffentlichten Bildern und Zeichnungen und natürlich allzeit gute Fahrt auf den Straßen dieser Welt.

GRUSSWORT
/
WORDS
OF WELCOME

A car that bears the name MINI is far more than just the sum of its parts, much more than just a consumer durable or an attractive mode of transport. As our customers confirm time and again, the MINI has its own quite particular character.

Every MINI inspires and brings people together all across the world. The brand transcends borders and sets trends. A relationship with a MINI is above all emotional. The slogan 'like a friend' is a declaration by thousands of fans all over the world, who do not want to go without a MINI in their life.

I too have a very personal relationship with this extraordinary brand. Many years ago, I and a lovingly restored classic Clubman tore up the streets of Madrid. I will never forget the experience, the feeling conveyed by my MINI, and the great deal of affection that others felt for my vehicle. As my companion the Clubman turned these trips into a very special adventure. There is of course one other link between the MINI and me – we were born the same year, in 1959.

For decades MINI has been creating the history of our industry – and it still is today. This illustrated book will immerse you in the history of the MINI. Be inspired by the different facets of this brand and relive its development page by page, from its beginnings all the way to the present. I hope you enjoy these photographs and drawings – some of which have never been published before – and I wish you happy motoring throughout the world.

Peter Schwarzenbauer
Mitglied des Vorstands
der BMW AG / Member of
the Board of Management
of BMW AG

Am Anfang einer Publikation, die der Marke MINI gewidmet ist, muss die Frage erlaubt sein, was einen MINI zum MINI macht. Schon zwei Generationen vor uns standen dem Mythos dieser Marke fasziniert gegenüber. Heute geht es uns nicht anders. Oft schon wurde der Versuch unternommen, mit rationaler Deutung oder blumigen Worten das Phänomen MINI zu fassen, Bemühungen, die meines Erachtens bisher nicht so recht überzeugen konnten. Im Gegenteil, je mehr man hier zu ergründen versucht, umso mehr Fragen werden aufgeworfen. Ist MINI mehr als ein Fahrzeug – am Ende gar ein Lebensgefühl? Oder ist es nicht mehr als ein kühl kalkuliertes Marketinggeschöpf? Es war mein Glück, dass ich die Marke MINI seit ihrem Neubeginn 2000 intensiv begleiten durfte – in beruflicher wie auch privater Hinsicht. Aus dieser Perspektive ist mir in dem Kleinwagen ein regelrechter Freund der Familie erwachsen. MINI ist eine ehrliche Haut, mitunter zickig, je nach Motorisierung aufgepumpt mit Testosteron oder gutmütig und »sweet«. Jeder MINI ist authentisch – eben in der Art, wie man es von einem guten Freund erwartet. Diesem Fahrzeug verzeiht man vieles, was man anderen nicht durchgehen ließe. Das merkt man dann, wenn man in einem MINI unterwegs ist und einen Fahrstil an den Tag legt, der nur »harakirimäßig« genannt werden kann. Der Verkehr hätte jeden anderen Raudi auf der automobilen Lichtung längst zur Strecke gebracht. Als MINI-Fahrer aber erlebt man Situationen, in denen einem zugelächelt und mit einem Handzeichen die Vorfahrt eingeräumt wird.

Warum ist das so? Liegt es am sogenannten »Kindchenschema«, das heißt an den Kulleraugen in Verbindung mit weiteren rundlich-kindlichen Formen, denen wir – archaisch geprägt – nicht widerstehen können? Liegt es an den lieb gewonnenen Design-Ikonen, die den Wagen stilistisch so einzigartig machen? Oder liegt es an der Klassenlosigkeit des MINI, der nachweislich Menschen verbindet und Grenzen überwindet? Am Ende vermute und hoffe ich, dass wir das Faszinosum MINI nie wirklich greifen, geschweige denn entmystifizieren werden. Um es mit Gertrude Stein zu sagen: Ein MINI ist ein MINI ist ein MINI. Haben sich schon Generationen an der Definition seines Wesens versucht – wir sollten ihn bewundernd so nehmen wie er ist.

VORWORT
/
FOREWORD

fine

Gabriele Fink
Leiterin BMW Museum,
München /
Director of the BMW Museum,
Munich

A publication dedicated to the MINI brand may surely start with the question, what makes a MINI a MINI in the first place. This legendary brand has already fascinated two previous generations. It's a real MINI! And it is no different for us today. Many attempts have been made to capture the phenomenon of the MINI through rational interpretations or flowery words. In my opinion none of these efforts has been entirely convincing from a holistic perspective. On the contrary, the more you try to discover its secrets, the more questions are raised. Is a MINI more than a car – perhaps even an attitude to life? Or is it nothing more than a cool, calculated marketing creation? I have been lucky enough to be able to witness up close the re-birth of the MINI brand that began in 2000 – both professionally and personally. From this perspective, this compact small car has for me become a real friend of the family. MINI is an honest soul, at times a little moody: depending on its level of mechanisation it's either pumped full of testosterone or sweet and docile. Every MINI is authentic, in the way that you would expect a good friend to be. We forgive this car much that we would not accept from others. You notice this when you are on the road in your MINI and find yourself driving in a way that can only be compared to *hara-kiri*. The cars behind you would have cut up any other hooligan on the road. But instead, MINI drivers earn a smile and a wave and are given right of way.

Why is this the case? Is it because of those 'child-like characteristics', those googly eyes combined with other round, child-like features to which we are instinctively drawn? Is it because of the design icons that we have come to love, which make the car unique? Or is it because of the classless appeal of the MINI, which has been shown to bring people together and to transcend boundaries? In the end, I can only assume and hope that we will never really understand the fascination of the MINI, let alone demystify it. As Gertrude Stein might say: A MINI is a MINI is a MINI. While generations have tried to define its nature, we should just adore the MINI and accept it for what it is.

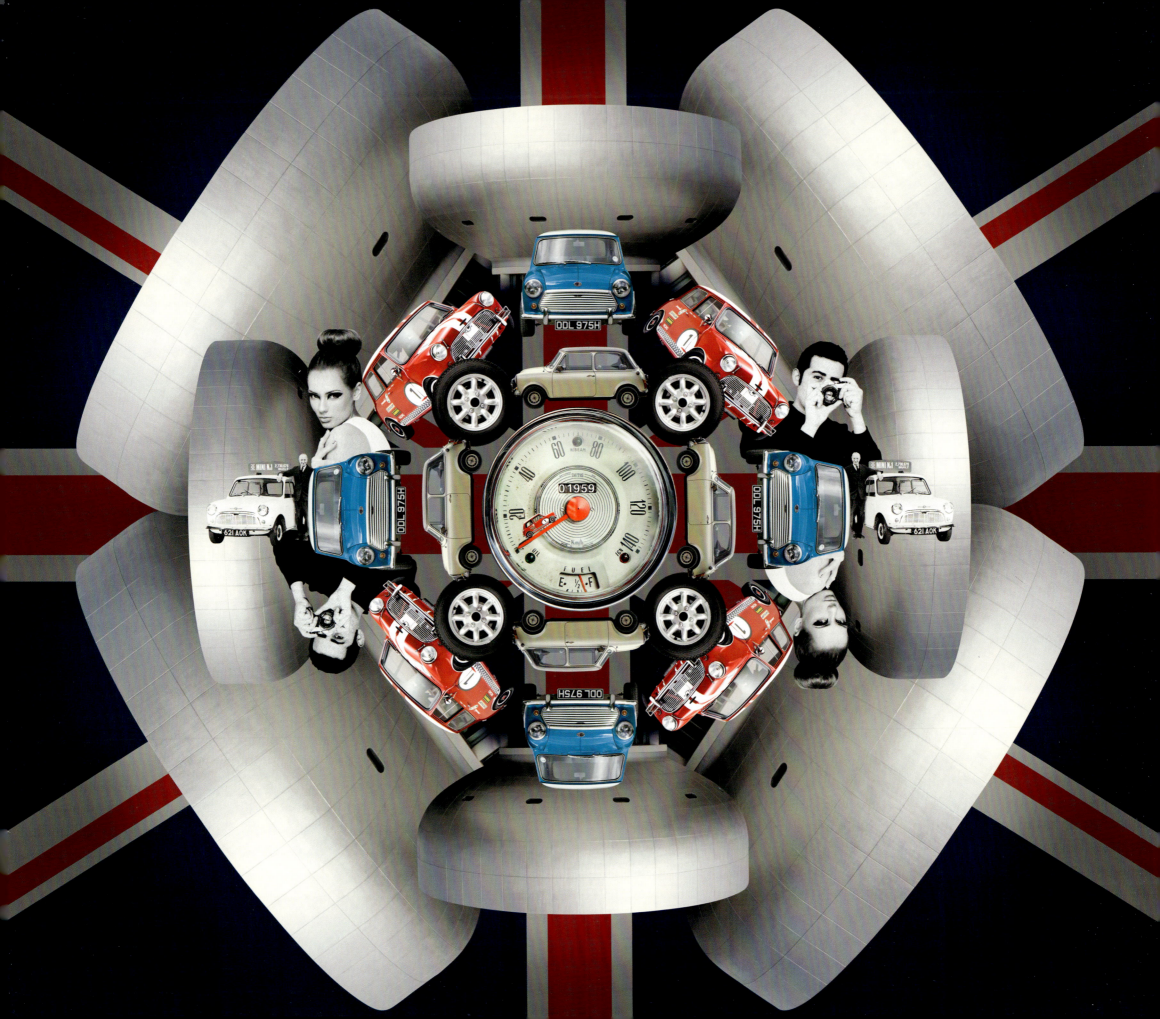

Das Wort »MINI« klingt cool und sympathisch. Ausschlaggebend dafür ist vermutlich die Silbendoppelung mit dem hellen Vokal. Die Bezeichnung, die im Englischen wurzelt, ist eine Kurzform von »Miniature« und meint das sehr Kurze, Kleine, Winzige. »MINI« ruft positive Assoziationen hervor, denn vom Kleinen geht per se das Gute aus. »MINI« genießt sozusagen »Welpenschutz«.

In Zusammenarbeit mit dem Hirmer Verlag ist ein Katalog entstanden, der sich der Marke MINI von ihren Anfängen bis heute widmet. Fahrzeugkonzepte, Designsprache, Produktion und gesellschaftliche Entwicklung verdienen eine genauere Betrachtung. Aus diesem Ansinnen hat sich eine Zeitreise entwickelt, die den Leser informieren, aber auch faszinieren will. Dem vorliegenden Buch geht die Gestaltung einer größeren Ausstellung im BMW Museum in München voraus, die über den Zeitraum von November 2014 bis Anfang 2016 zwanzig übergreifende Themenbereiche vorstellt. Wir freuen uns, dass es uns in diesem Kontext gelungen ist, mehr als dreißig Originalfahrzeuge zusammenzutragen und diese gemeinsam mit seltenen Handzeichnungen sowie kostbaren Kleinexponaten präsentieren zu können.

Herzlich danken möchte ich allen, die zum Gelingen dieser großen Aufgabe beigetragen haben, allen voran dem British Motor Industry Heritage Trust in Gaydon, der mir seine Archive öffnete. Wertvolle Designbeiträge lieferten darüberhinaus das Victoria & Albert Museum in London sowie die Neue Sammlung der Pinakothek der Moderne in München. Ferner erhielt ich freundliche Leihgaben von mehreren Firmen, Privatsammlern und Werkstätten. Auch die freundliche Unterstützung von Samsung Electronics verdient an dieser Stelle Erwähnung. Mein persönlicher Dank gilt ebenso den vielen MINI-Experten, die mit Rat und Tat zur Seite standen, vor allem Tanya und Jason Field, Wouter de Graaf, Keith Adams und Moritz Burmester. Schließlich möchte ich allen Kolleginnen und Kollegen der BMW Group meinen Dank aussprechen, die in den Bereichen Kommunikation und Produktion, Archiv und Marketing wertvolle Beiträge leisteten.

Bei der Entwicklung des Konzepts für Katalog und Ausstellung stand zunächst der Wunsch im Mittelpunkt, die Geschichte von MINI

EINFÜHRUNG
/
INTRODUCTION

The word 'MINI' sounds both cool and attractive. This is probably a result of its two syllables and front vowels. The name is an abbreviation of the English word 'miniature', meaning something very short, small or tiny. 'MINI' triggers positive associations because small in itself means good, and thus 'MINI' benefits from our instinct to protect the small.

This catalogue devoted to the MINI brand from its beginnings to the present day was developed in collaboration with Hirmer Verlag. The concepts, language of design, production and social trends associated with the MINI all deserve close examination. With this in mind, we embarked upon a historical journey, whose aim is to inform, but also to inspire, the reader. This book is based on the design of a larger exhibition comprising 20 overarching thematic sections, which is showing at the BMW Museum in Munich from November 2014 to early 2016. On this occasion we are delighted to have been able to assemble over 30 original vehicles, and to exhibit them alongside rare hand drawings and construction sketches as well as precious small items from the world of the MINI.

I would like thank all those who have contributed to the success of this great task, above all the British Motor Industry Heritage Trust in Gaydon, which made its archives available to me. The Victoria & Albert Museum in London and Die Neue Sammlung der Pinakothek der Moderne in Munich also made valuable design contributions. In addition several companies, private collectors and workshops kindly lent items to the exhibition. The generous support of Samsung Electronics deserves special mention. I would personally also like to thank the many MINI experts who offered both advice and resources, especially Tanya and Jason Field, Wouter de Graaf, Keith Adams and Moritz Burmester. Finally, I would like to thank all my colleagues at BMW Group, who made valuable contributions in the areas of communication and production, archives and marketing.

In developing the concept for both catalogue and exhibition, the main aim was to present the history of the MINI as a whole, and to highlight the famous guiding thread that has remained unbroken from

in ihrer Gesamtheit darzustellen, den berühmten »roten Faden« aufzuzeigen, der seit 1959 ohne Unterbrechung bis heute gesponnen wurde. Das Gemeinsame, nicht das Trennende gab den Impuls für dieses Projekt. Dass Marken und Unternehmen sich verändern, dass sie ihre Produkte den Erfordernissen der Zeit anpassen, ist selbstverständlich. Dennoch ging für manchen Zeitgenossen mit dem Ende des klassischen Mini eine Ära zu Ende. Aus nostalgischer Sicht kann man diese Wehmut verstehen. Das für MINI bedeutende Jahr 2001 bescherte der Marke jedoch eine wichtige Kurskorrektur, denn sie bedeutete eine dringend erforderliche Weichenstellung, dank derer sie heute zuversichtlich in die Zukunft blickt. Natürlich baut der MINI heutiger Prägung auf den großen Erfolgen der Vergangenheit auf und bekennt sich zur eigenen Historie, er gibt aber auch etwas Wesentliches zurück, belebt er doch die klassische Zeit vor 2001.

Spätestens hier stellt sich die Frage nach der Schreibweise des Namens. Gern hätte der Kurator im Zuge der Harmonisierung eine einheitliche »Wortmarke« gewählt, doch sprechen die historische Gepflogenheit und markenrechtliche Vorgaben dagegen. So ist im Folgenden vom classic Mini oder Mini die Rede, wenn die Ära 1959 bis 2000 gemeint ist, während sich MINI ab 2001 in Versalien schreibt. Ist das Phänomen der Marke in ihrer Gesamtheit angesprochen, soll hier die Vereinbarung gelten, MINI ebenfalls groß zu schreiben.

Die Marke MINI, deren Fundament 1959 gelegt wurde, hat sich in den folgenden 55 Jahren weiterentwickelt. Voller Stolz und Respekt blicken die heute Verantwortlichen auf die lange Tradition zurück. Das ist kein bloßes Lippenbekenntnis, sondern wird durch die Beibehaltung von Proportionen, Designmerkmalen, Grundprinzipien des Fahrzeugaufbaus und Modellbezeichnungen evident. Bei aller Differenzierung steht die Kontinuität der Identität von MINI im Vordergrund. Was die Ära vor 2001 mit der heutigen verbindet, ist der Ausdruck von Sympathie, Leidenschaft und Lebensfreude.

Allein schon das Äußere des MINI weckt die Sympathie des Betrachters. Die verhältnismäßig großen Scheinwerfer, die hoch gezogene Windschutzscheibe und die kleinen Räder tragen zu dem

1959 to the present day. Common features rather than differences were the driving force of this project. Of course brands and companies change, and they adapt their products to the requirements of the age. Nevertheless, for many people at the time the end of the classic Mini was also the end of an era. From a nostalgic perspective this sense of regret is understandable. However in 2001 – that all-important year for the MINI – the brand underwent an important change of course, as it urgently needed a new direction, thanks to which it can now be confident about the future. Naturally, today's MINI is based on the great successes of the past and recognizes its own history. But it also gives something important back by breathing new life into the classic, pre-2001 era.

It is at this point, if not before, that the question arises of how the name should be written. For the sake of harmonisation the curator would have preferred one word as a consistent brand, but historic practice and trademark law prevent this. As a result, in this book we talk about the classic Mini or Mini whenever the period between 1959 and 2000 is meant; while the post-2001 MINI is written in capitals. Whenever the brand in general is discussed, it has been agreed that in this instance, MINI will appear in capitals.

The MINI brand, whose foundations were laid in 1959, has developed enormously over the subsequent 55 years. Today's management look back on the long tradition with pride and respect. This is not merely paying lip service, but is evident in the decision to maintain proportions, design features, and the basic principles of the car structure and model names. Despite many differences the focus remains on the continuity of the MINI's identity. What connects the pre-2001 period with today is the MINI's ability to express likeability, passion and *joie de vivre*.

In itself the MINI's exterior is appealing. The relatively large headlights, high windscreen and small wheels further contribute to what psychologists mights call the key stimuli of childlike characteristics.

The original 1959 Mini was the first compact car that was not only tiny but also had all the attributes of a 'normal' car. It was a genuine

bei, was Psychologen als den Schlüsselreiz eines Kindchenschemas bezeichnen.

Der Ur-Mini von 1959 war der erste Kleinwagen, der nicht einfach nur winzig war, sondern zugleich alle Zutaten eines »normalen« Autos besaß. Er war die echte Miniaturausgabe eines ausgewachsenen Automobils. Gern wird an dieser Stelle der Vergleich zwischen David und Goliath gezogen. Der kleine Mini gegen die vermeintliche Übermacht großer Konkurrenten, so wie im Alten Testament der hebräische David den fast 3 Meter großen, schwer gepanzerten Philister Goliath mit einem Stein bezwang. Das Entscheidende an dieser Episode wird oft übersehen: David war so clever, seine primitive Steinschleuder als Fernwaffe einzusetzen. Um gegenüber dem Größeren zu bestehen, muss man nicht unbedingt größer oder stärker, aber intelligenter sein. Genau das ist einer der Wesenszüge des MINI: Sein Raumkonzept – ob von 1959 oder 2001 – ist nahezu genial. Auf sehr begrenzter Grundfläche bietet der MINI ein Höchstmaß an Volumen, was der Slogan »clever use of space« treffend umschreibt. Unter den Kleinen ist er der Größte, unter den Großen zählt er zu den Kleinsten.

Der MINI ist von Beginn an cool. Man liebt seinen unkonventionellen Auftritt, seinen Charakter, sein Anderssein. Das Spießige ist ihm ebenso fremd wie das stillose Protzen. Vielmehr gibt er sich bescheiden bzw. trägt er das Schwierige und Komplexe mit ostentativer Lässigkeit vor. MINI steht für das Luftige und Leichte, für eine bestimmte Lebenseinstellung. MINI hat eine britische Heimat, ist ansonsten aber nationenlos, klassenlos, kompatibel, individuell und zugleich offen und vernetzt.

Fast über Nacht war der Kleinwagen zum Kultobjekt geworden. Sein innovativer und nonkonformistischer Charakter passte perfekt zum Zeitgeist der Swinging Sixties. Er war ein Auto, das sich von allem anderen unterschied, gegen vorherrschende Konventionen verstieß und obendrein eine ganze Menge Fahrspaß bereitete. Bis heute hat es die Marke verstanden, jung, dynamisch und aktuell zu bleiben. Nur wenige Fahrzeugkonzepte haben ähnlich lange Zeitspannen überdauert oder eine vergleichbare Popularität erlangt.

miniature version of a fully grown automobile. At this point the comparison between David and Goliath is often raised – the small Mini against the apparent superiority of its larger rivals, just as the Israelite David in the Old Testament used one stone to defeat the Philistine Goliath who was almost 3 m tall and fully armoured. The key point of this episode is often overlooked: David was clever enough to use his primitive sling-shot as a long-range weapon. To prevail against a bigger opponent, size is less important than superior intelligence. And this is one of the MINI's essential features: the way it uses space – whether in 1959 or in 2001 – is a stroke of genius. The MINI offers maximum volume on a very limited footprint, perfectly described by the slogan 'clever use of space'. It is the biggest among the small, and the smallest among the big.

The MINI has been cool right from the start. It is loved for its unconventional looks, its personality and the fact that it is different. It is as unlikely to be straight-laced as it would engage in tasteless posturing. Rather, it comes across as modest, expressing difficult and complex ideas with pointed nonchalance. MINI represents all that is light and airy, and a special approach to life. MINI's home is in Britain but otherwise it is nationless, classless, adaptable, individual and simultaneously open and connected.

This compact car became a cult object almost overnight. Its innovative and non-conformist character was a perfect match for the spirit of the Swinging Sixties. This car was different from all others, transgressing dominant conventions and offering a whole lot of driving fun into the bargain. To this day, the brand has managed to stay young, dynamic and up-to-date. Only a small number of other cars has survived for as long a period of time or enjoyed similar popularity.

The revolutionary compact car of 1959 became a classic in the history of cars, a timeless automobile that readers of British magazine *Autocar* voted Car of the Century in 1995.

In retrospect, the MINI's success story has been largely down to chance. We recall that to begin with the original target group of buyers, British middle-class families, were not interested in this compact,

Aus dem revolutionären Kleinwagen von 1959 wurde ein Klassiker der Automobilgeschichte, ein zeitloses Automobil, das 1995 von den Lesern der britischen Zeitschrift *Autocar* zum »Auto des Jahrhunderts« gewählt wurde.

Dabei verdankt sich die Erfolgsgeschichte von MINI rückblickend vor allem dem Zufall. Es sei daran erinnert, dass die ursprünglich angepeilte Käuferschicht, die Familien in Großbritanniens Mittelschicht, zunächst keinen Gefallen am kompakten, spritsparenden Kleinwagen fand. Zu unkonventionell war ihnen das Design. Stattdessen interessierte sich dafür bald eine andere Zielgruppe – jene, die sich einen Zweitwagen leisten konnte. In Anlehnung an die *Mythen des Alltags* von Roland Barthes (1957) ist ein Auto wie der MINI aufgetaucht, wie ein magisches Objekt, als wäre es vom Himmel gefallen. Er ist mehr als ein vergängliches Produkt unserer Konsumwelt, eine große epochale Schöpfung.

Auch der Erfolg des neuen MINI ab 2001 ist nicht allein das Produkt kalkulierbarer Marketingstrategien, sondern baut vor allem auf der Kreativität und Leidenschaft derer auf, die ihn entwickeln und permanent neu denken. All denen, die ihn schätzen, besitzen, fahren und pflegen, die sich für den Mini wie den MINI begeistern, ist dieses Buch gewidmet.

fuel-saving small car, as they found its design too unconventional. Instead, it soon attracted the interest of another target group – those who could afford to buy a second car. As Roland Barthes might have said in his *Mythologies* (1957), a car like the MINI appeared like a magical object, as if it had fallen from the skies. It is more than a transient product of our consumer society; it is a major, epoch-making creation.

The success of the new, post-2001 MINI is not just a result of calculated marketing strategies either, but builds primarily on the creativity and passion of those who develop it and who constantly re-imagine it. This book is dedicated to all those who appreciate, own, drive, look after and are enthusiastic about both the Mini and the MINI.

Andreas Braun

Dr. Andreas Braun
Kurator BMW Museum,
München / Curator BMW
Museum, Munich

IDENTITÄT UND LIFESTYLE
/
IDENTITY AND LIFESTYLE

Bereits beim ersten Mini, der 1959 der Öffentlichkeit vorgestellt wurde, setzte die Werbung auf den besonderen Charakter des kleinen Raumwunders. Die frühen Anzeigen sind betont farbenfroh und illustrieren die Vorzüge des revolutionären Kleinwagens. Glücklich strahlende Gesichter vierköpfiger Familien oder entsprechend vieler Erwachsener genießen die Fahrt im Mini, bewundern das enorme Platzangebot und nutzen die ausgeprägte Funktionalität. Schon die frühe Printwerbung bewies Sinn für Humor und bot manchen Slogan eher mit einem Augenzwinkern. So wurde dem Morris Mini-Minor attestiert, er sei das »aufregendste Auto der Welt«.

Heutige Anzeigen der Marke MINI sprechen selbstverständlich eine andere Sprache. Sie fallen durch ein plakatives Arrangement auf, das den Fokus auf das neue Fahrzeug lenkt und ihm einen schwarzen Hintergrund gibt – aufgelockert durch einen bunt leuchtenden Rahmen und eine auf ein Minimum reduzierte Überschrift. Die Tonalität orientiert sich an den modernen Zielgruppen und hebt sich deutlich vom Wettbewerb ab. An der Sprache der Werbung wird deutlich, dass MINI die am stärksten emotional profilierte Marke im Kleinwagenbereich ist.

Früh wurde bei MINI das Potenzial des Fernsehens genutzt. Für unterschiedliche Märkte kreierte man spezielle TV-Spots, in die der kulturelle Background des Publikums einbezogen wurde. Ob als perfekte Lösung im Pariser Verkehrsgewühl oder als ideales Transportmittel in einer Wüstengegend Australiens – immer präsentierte sich der Mini als das richtige Auto am richtigen Ort. Auch im Zeichentrickfilm stellte er humorvoll und selbstironisch seine Überlegenheit unter Beweis.

Der neue MINI stieß auf Anhieb weltweit auf großes Interesse. Standen zu Beginn die Marke und der MINI Hatch im Vordergrund, sind es heute eher einzelne Modelle, deren Positionierung vermittelt wird. Hierbei werden ausgewählte Bezüge zur reichen Historie des MINI geschaffen. Und das nicht ohne Grund, schließlich stellt der MINI ab 2001 eine Weiterentwicklung des klassischen Mini dar.

Da es in der Werbung nicht allein um Verkaufsförderung geht, sondern ebenso um die Schaffung eines Markengefühls, ist Kreativität

MINI IN DER WERBUNG / MINI IN ADVERTISING

Advertisements for the first Mini, which was introduced to the public in 1959, focused from the start on the particular character of this small miracle of space. Early ads are highly colourful and illustrate the advantages of the revolutionary small car. The happy, smiling faces of a family of four or four adults are depicted enjoying a Mini, admiring the enormous amount of available space and making the most of its excellent functionality. A sense of humour was evident even in the early print ads, which incorporated a nod and a wink into some of the advertising slogans, such as the claim that the Morris Mini-Minor, was the 'Most Exciting Car in the World'.

Today, of course, advertising for the MINI brand speaks another language. It attracts attention through bold graphic designs that focus on the new vehicle set against a black background, broken up by a bright colourful frame and a minimal headline. The colour palette is oriented towards modern target groups and differentiates itself clearly from the competition. This advertising language makes it clear that MINI is the brand with the most powerful emotional profile in the small car sector.

The potential of television was used in MINI advertising right from the start. Special TV commercials were created for different markets, incorporating the cultural background of the audience. Whether as the perfect solution to the turmoil of Paris traffic, or as an ideal means of transportation in the Australian desert, the Mini was always shown to be the right car in the right place. In cartoons, the Mini proved its superiority through humour and self-irony.

The new MINI attracted huge international interest from the start. While the brand itself and the MINI Hatch initially took centre stage today the focus is on individual models and their positioning. Such advertising makes use of selected references to the MINI's rich history – and not without reason as after all the post-2001 MINI represents a further development of the classic Mini.

As advertising is not simply about increasing sales but just as much about the creation of a feeling associated with the brand, creativity is important. MINI marketing has won several international prizes over the last few years for its inventive TV commercials.

gefragt. Für seine originellen TV-Spots wurde das MINI Marketing in den letzten Jahren vielfach mit internationalen Preisen ausgezeichnet.

Neben klassischer Kommunikation setzte MINI immer schon auf andere innovative Formen wie das sogenannte Guerilla-Marketing und nutzt damit den hohen Aufmerksamkeitswert ungewöhnlicher Aktionen. Originelle Auftritte, die oft nur für kurze Zeit im öffentlichen Raum wahrgenommen werden, verblüffen die Passanten. Frech, keck, aber auch charmant realisiert der MINI scheinbar Unmögliches und schafft es stets, seine Botschaften abzusetzen bzw. Spuren zu hinterlassen. Seit dem Relaunch der Marke 2001 wurden bereits rund 200 maßstabsgetreue MINI-Modelle aus Fiberglas angefertigt und weltweit an Orten eingesetzt, wo man sie nie vermutet hätte – an steil aufragenden Hauswänden, auf Kaminschloten, Dächern oder unter Brücken. Im Ankunftsbereich eines Flughafenterminals schmückt die Heckansicht des MINI Clubman die Ausgangstüren. Unweit eines Messegeländes ziehen vier aufrecht stehende, bis zur Windschutzscheibe im Boden vergrabene MINI-Fahrzeuge die Aufmerksamkeit auf sich. In Kanada wurde der MINI sogar auf einer Plakatwand in ein gespanntes Katapult gezwängt. Spektakulär sind vor allem die Aktionen, die die Möglichkeit von Handy und Internet nutzen: Per Anruf vom Handy konnten Passanten 2006 ein Plakat, das den MINI Cooper S zeigte, zum Leben erwecken. Unter dem Motto »Wecken Sie den Stier« war es möglich, den Ausstoß von Rauchwolken aus den »Nüstern« des MINI zu aktivieren.

Der neue MINI-Fünftürer / The new five-door MINI, 2014

AUSTIN mini countryman
For business or pleasure – dual purpose perfection

MORRIS MINI-MINORS
TODAY'S CAR IS A MINI

IT'S WIZARDRY ON WHEELS!
The Revolutionary "QUALITY FIRST" **MORRIS** Mini-Minor

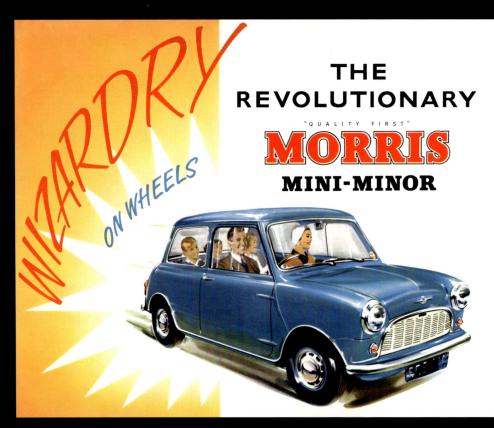

WIZARDRY ON WHEELS
THE REVOLUTIONARY "QUALITY FIRST" **MORRIS** MINI-MINOR

AUSTIN Incredible **mini** Saloon . . . .

★ Combined ignition/starter switch.
★ Safety sun visors and interior mirror.
★ Two-leading-shoe brakes at front.
★ Greater torque capacity gearbox.

. . . now with Hydrolastic suspension!

AUSTIN MINI-COOPER 1000 1275 'S' TYPES

AUSTIN seven *Countryman*

BUSINESS BEFORE PLEASURE . . .

GREAT LITTLE CARS

MORRIS MINI-MINORS

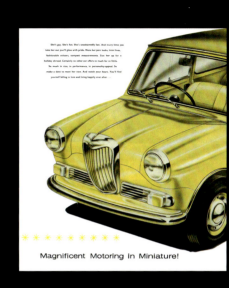

Magnificent Motoring in Miniature!

Wizardry at work again !

QUALITY FIRST
THE **MORRIS** MINI-TRAVELLER

Double-up with the dual purpose...

AUSTIN
INCREDIBLE **mini** SALOON & COUNTRYMAN

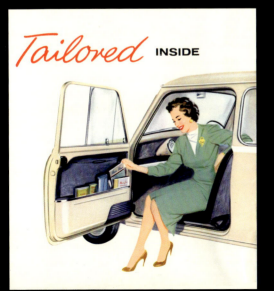

Tailored INSIDE

AUSTIN **mini** *Countryman*

dual-purpose perfection

BMC

Mini Austin Seven, 1959

Alongside classic forms of advertising, MINI has always employed other innovative methods such as so-called 'guerrilla marketing', thus exploiting the high attention value associated with unusual promotions. Ingenious events that are only on public display for a short period can take people by surprise. Impudent, bold, but also charming, MINI realizes the seemingly impossible and always manages to communicate its message and make a lasting impression. Since the relaunch of the brand in 2001, approximately 200 full-size, fibreglass model MINIs have been made and installed across the world in completely unexpected locations – on the sides of tall buildings, on chimneystacks, on roofs or under bridges. In the arrivals area of an airport terminal, the rear end of a MINI Clubman adorns the exit doors. Near an exhibition centre, four MINIs buried upright in the ground up to their windscreens create an attention-grabbing sight. In Canada, a MINI was even placed in a stretched catapult on an advertising hoarding. Most spectacular of all are promotions that make use of mobile phones and the Internet. In 2006 passers-by were able to make a poster of the MINI Cooper S come to life by calling from their mobile phone. In response to the slogan 'Wake the Bull', viewers could activate clouds of smoke blowing out of the MINIs 'nostrils'.

MINI AT THE OLYMPICS

Gern denken wir an die groß-artige Olympischen Sommer-spiele in London 2012 zurück, an die Erfolge der Athleten, die im Rampenlicht standen, an glückliche Siege und bittere Niederlagen. In Erinnerung bleiben auch eine faszinierende und großartige Eröffnungs- und Schlussveranstaltung. Und da ist jenes Gefährt, das im Olympiastadion fern-gesteuert langsam über den Rasen fuhr und das Publikum vor Ort wie auch weltweit vor dem Fernseher zum Schmunzeln brachte.

Es war einer dieser MINI-typischen Marketing-Coups: Speziell für die Olympischen und Paraolympischen Spiele

wurden drei MINI Hatch im Kleinformat im Maßstab 1:4 entwickelt. Sie hatten die Auf-gabe, den Athleten das Leben während der Leichtathletik-wettbewerbe zu erleichtern. Unter ihrem abnehmbaren Sonnendach befand sich ein Stauraum für diverse Utensilien: In den Disziplinen Speer-, Dis-kus- und Hammerwerfen sowie Kugelstoßen standen sie rund um die Uhr bereit und trans-portierten die abgelegten Sportgeräte vom Feld zurück in die Wurfbereich. Das sparte wertvolle Zeit im Ablauf der Wettbewerbe. Jeder dieser sogenannten Mini MINI legte während seiner täglichen Vier-Stunden-Schicht etwa 6 Kilo-

meter zurück und schleppte dabei bis zu 8 Kilogramm Ge-wicht durch die Gegend. Als Antrieb diente ein 10 PS starker, abgasfreier Elektromotor, der von eigens ausgebildeten »GameMakern« per Funk-steuerung gelenkt wurde. Der dienstbare Helfer mit Spezial-bereifung wurde aus einem Leichtbau-Chassis aus Ver-bundwerkstoff gefertigt und verblüfft beim näheren Hin-sehen durch originalgetreue Details wie Türgriffe, Außen-spiegel, Scheibenwischer und funktionierende Scheinwerfer. Es ist nicht auszuschließen, dass er bei zukünftigen sport-lichen Großereignissen wieder zum Einsatz kommt.

It is enjoyable to look back on the magnificent summer Olympic Games held in London in 2012, the achievements of the athletes in the limelight, the happy victories and bitter defeats. The superb, fascinat-ing opening and closing cere-monies have also created last-ing memories. And above all we remember the car that en-tered the Olympic stadium by remote control, drove slowly across the grass and raised a smile from both the audience in the stadium and those watching on television across the world.

It was a typical MINI mar-keting coup. Three MINI Hatch models in a smaller 1:4-scale

were developed especially for the Olympic and Paralympic Games. Their job was to make the athletes' lives easier dur-ing the track and field competi-tions. Beneath its removable sun roof was storage space for various items, as the cars were available round the clock for the javelin, discus and hammer competitions as well as shot-put, carrying discarded sport equipment off the field and back to the throwing area, and thus saving valuable time dur-ing competitions. Each of these so-called 'Mini MINIs' travelled about 6 km during its daily 4-hour shift, transporting loads of up to 8 kg across the park. The 10 bhp, zero-emissions

electric motor was radio-con-trolled by specially trained Games Makers. These little helpers boasted special tyres and a lightweight chassis made of composite material. Astonishingly, a closer look revealed original details such as door handles, side mirrors, windscreen wipers and func-tioning headlights. It is likely that they will be used again in the future at other major sporting events.

In zahlreichen TV- und Kinofilmen hat der MINI bisher als Komparse bzw. Hauptfigur mitgewirkt. Im Kultstreifen *Blow up* von 1966 ist er ebenso zu sehen wie in den Comedy-Streifen von *Mr. Bean*.

Der wohl berühmteste Film zu diesem Thema ist *The Italian Job* von 1969. Die Gangsterkomödie wurde zum Kinoerfolg und Kultstreifen der Sechzigerjahre. Gemeinsam mit seinen Komplizen plant der britische Gauner Charlie Croker, gespielt von Michael Caine, einen Goldraub mitten in Turin. Zunächst sollen die italienische Polizei und die örtliche Mafia durch ein Verkehrschaos irritiert werden, damit die verwegenen Diebe in drei Mini Cooper mit dem Gold entkommen können. Das Ganze mündet jedoch in eine abenteuerliche Hetzjagd über marmorne Freitreppen, durch die Kanalisation und über das Dächermeer der Stadt, wobei sich die kleinen Autos als ideale Fluchthelfer erweisen. Die spektakulären Verfolgungsrennen mit dem roten, blauen und weißen Mini schrieben Filmgeschichte.

Auch das Remake *The Italian Job* von 2003 wartet mit ähnlich spannenden Fahrszenen auf und wurde ebenso zum großen Erfolg. In der Neuverfilmung mit den Hollywoodstars Charlize Theron und Mark Wahlberg wurde die Geschichte um einen spektakulären Goldraub noch effektvoller in Szene gesetzt. Der Action-Thriller beginnt mit einer Verfolgungsjagd durch die engen Wasserstraßen von Venedig. Im Lauf des Films erbeuten die Gauner Gold im Wert von 35 Millionen Dollar. Am Ende stellt sich die bange Frage, ob sie die Beute in und unter den Straßen von Los Angeles auch behalten. Ungeachtet der Gewissheit, dass es sich um verlustreiche Dreharbeiten handeln würde, stellte MINI 32 Fahrzeuge des MINI Cooper S zur Verfügung. Weil die Rennen im U-Bahnbereich keine Fahrzeuge mit Benzinmotoren erlaubten, wurden eigens drei MINI mit Elektroantrieb ausgestattet.

Der prominente britische Schauspieler Peter Sellers brachte die Filmkarriere des Mini in Gang, als er ihn 1964 für den Film *A Shot in the Dark (Ein Schuss im Dunkeln)* engagierte, die Fortsetzung der legendären Gaunerkomödie *Der rosarote Panther*. Als Inspektor Clouseau untersucht Sellers den Mordfall in einer reichen Familie. Gemeinsam mit dem Zimmermädchen Maria alias Elke Sommer, das des Mordes

MINI IM FILM
/
MINI AT THE MOVIES

The MINI has appeared in numerous TV and cinema films both as an extra and as the main character. It features in *Blow-Up*, the 1966 cult movie, and in the comedy *Mr. Bean*.

But the Mini's most famous appearance must surely be in the 1969 movie *The Italian Job*. This heist comedy became a hit and a cult film of the 1960s. British crook Charlie Croker, played by Michael Caine, is plotting a gold robbery in the middle of Turin, aided by his accomplices. The plan is to thwart the Italian police and local Mafia by causing traffic chaos so that the audacious thieves can escape with the gold in three Mini Coopers. However, the whole event turns into an exciting chase down marble steps, through the sewer system and over roofs of the city with the small cars proving to be the ideal getaway vehicles. The spectacular chase scene featuring red, blue and white Minis became a classic of film history.

The 2003 re-make of *The Italian Job* provided similarly exciting chase scenes and enjoyed the same degree of success. In the new version with Hollywood stars Charlize Theron and Mark Wahlberg, the story of a spectacular gold robbery was brought to the screen even more effectively. The action thriller begins with a chase through the narrow waterways of Venice. The beginning of the film shows the crooks stealing gold to the value of $35 million. By the end, they are anxiously trying to hang on to their spoils on – and under – the streets of Los Angeles. Regardless of the certainty that it would be a loss-making venture, MINI provided 32 MINI Cooper S for the shoot. And as the Los Angeles subway system did not permit petrol-driven motors, three MINIs were specially fitted with electric engines.

The prominent British actor Peter Sellers launched the Mini's film career when he used it in 1964 in *A Shot in the Dark*, the sequel to the legendary caper movie *The Pink Panther*. In the role of Inspector Clouseau, Sellers is investigating a case of murder in a wealthy family. Together with chambermaid Maria, played by Elke Sommer, who is suspected of the murder of her boyfriend, he embarks upon a journey in a Mini Cooper that is all the more embarrassing because both of them are naked.

In October 1965 *The Avengers* TV series hit the screens for the first time. The stories revolve around a mismatched pair of secret agents who solve the most complicated criminal cases. Gentleman John Steed, played by Patrick Macnee, maintains close contacts with the military and is a serious wine connoisseur. Alongside him, Emma Peel, played by Diana Rigg, upholds law and order. She embodies the fashion-conscious emancipated woman who publishes in scientific periodicals and is a mistress of far-eastern martial arts. Both frequently make use of a Mini Moke to get themselves out of a tight spot.

The 1966 film *Two Weeks in September*, with Brigitte Bardot and Laurent Terzieff in the leading roles, features a remarkable scene in Paris when models in miniskirts drive around in Mini Coopers.

The Mini is also an obligatory character in several James Bond movies, such as the 1973 blockbuster *Live and Let Die* that became famous for its spectacular motorboat scenes. In his first outing as James Bond, Roger Moore wages a bitter battle against gangster boss Kananga and his Voodoo tricks. In one beach scene, so as not to lose any time, he jumps into a Mini Moke and takes up the pursuit without further ado.

In the 1981 action drama *Goodbye Pork Pie*, produced in New Zealand, Gerry Austin happens upon a wallet containing a driver's licence and immediately hires a Mini. After picking up two hitch-hikers, an exciting road movie with action-packed chases ensues. The worn-out Mini is made to suffer, as more and more vehicle parts must be sold owing to a permanent lack of funds.

The comedy series *Mr. Bean*, produced between 1990 to 1995, are as much a part of England as five o'clock tea, Big Ben, red letterboxes and Her Majesty the Queen. In 14 episodes of the prize-winning TV-series, the extremely uptight and eccentric main character – inimitably played by Rowan Atkinson – encounters all kinds of hilarious everyday situations where he cannot help putting his foot in it. His permanent companion is a 1977 green, yellow and black Mini Cooper. To prevent the car being stolen, he secures the driver's door with a padlock and removes the steering wheel as a precaution. Mr. Bean expects a lot

23

an ihrem Freund verdächtigt wird, unternimmt er in einem Mini Cooper eine Fahrt, die prekär wird, weil beide nackt sind.

Im Oktober 1965 wurde erstmals die TV-Serie *The Avengers (Mit Schirm, Charme und Melone)* ausgestrahlt. Im Mittelpunkt steht ein ungleiches Agentenpärchen, das komplizierteste Kriminalfälle löst. Da ist der Gentleman John Steed, gespielt von Patrick Macnee, der beste Kontakte zum Militär unterhält und sich als profunder Weinkenner zu erkennen gibt. An seiner Seite sorgt Emma Peel, eigentlich Diana Rigg, für Recht und Ordnung. Sie verkörpert die modebewusste, emanzipierte Frau, die in wissenschaftlichen Zeitschriften publiziert und fernöstliche Kampftechniken beherrscht. Mehrfach steht den beiden ein Mini Moke helfend zur Seite.

Bei dem Film *A Cœur Joie (Zwei Wochen im September)* von 1966 – mit Brigitte Bardot und Laurent Terzieff in den Hauptrollen – kommt es in Paris zu einer bemerkenswerten Szene, als Mannequins in Mini-Röcken Mini Cooper fahren.

Auch in mehreren James-Bond-Streifen darf der Mini nicht fehlen, so in dem Kinoschlager *Live and Let Die (Leben und Leben lassen)* von 1973, der für seine spektakulären Motorboot-Szenen berühmt wurde. In seiner ersten Rolle als James Bond führt Roger Moore einen erbitterten Kampf gegen den Gangsterboss Kananga und die Tücken des Voodoo-Kults. Um keine Zeit zu verlieren, springt er bei einer Szene am Sandstrand kurzerhand in einen Mini Moke und nimmt die Verfolgung auf. Bei dem in Neuseeland 1981 produzierten Actiondrama *Goodbye Pork Pie (Ein Mini hängt die Bullen ab)* findet Gerry Austin zufällig eine Geldbörse mit Führerschein und leiht sich umgehend einen Mini aus. Nachdem sich zwei Anhalter dazugesellt haben, beginnt ein spannendes Roadmovie mit rasanten Verfolgungsfahrten. Der arg strapazierte Mini muss Haare lassen, denn da permanent das nötige Kleingeld ausgeht, müssen immer mehr Teile vom Fahrzeug verkauft werden.

Die 1990 bis 1995 gedrehten Episoden des *Mr. Bean* gehören zu England wie der Fünf-Uhr-Tee, Big Ben, rote Briefkästen und Ihre Majestät, die Queen. In den 14 Folgen der preisgekrönten TV-Serie begibt sich die überaus spießige und kauzige Hauptfigur – unnachahmlich

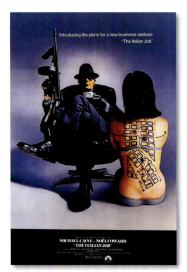 

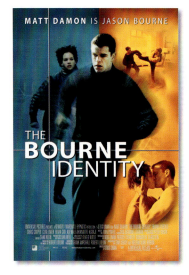 

MINI in der Filmgeschichte / MINI in the history of film

gespielt von Rowan Atkinson – in allerlei urkomische Alltagssituationen, ohne dabei ein Fettnäpfchen auszulassen. Sein ständiger Begleiter ist ein grüngelb und schwarz lackierter Mini Cooper von 1977. Damit ihm das Auto niemand klaut, sichert er die Fahrertür mit einem Vorhängeschloss und nimmt sicherheitshalber das Lenkrad mit. Vieles mutet Mr. Bean seinem Auto zu: Wenn er auf dem Dach einen Fernseher, ein anderes Mal einen Fernsehsessel, ja sogar mal einen 4 Meter hohen Weihnachtsbaum transportiert, wird der Kleinwagen kurzerhand zum Großtransporter erklärt. Unvergessen die Szene, in der der Mini Cooper unbemerkt vom daneben stehenden Besitzer von einem Panzer überrollt wird.

Eine postmoderne Parodie auf Großbritanniens Spionagefilme der Sechzigerjahre ist die Trilogie rund um *Austin Powers*. Mit dem dritten Teil *Austin Powers in Goldmember* kam 2002 erstmals auch der neue MINI ins Spiel. Drehbuchautor und Hauptdarsteller Mike Myers hatte für dieses Kinospektakel ein außergewöhnliches Aufgebot an Hollywoodstars sowie sechs MINI Cooper im Union-Jack-Dekor engagiert. In Anspielung an liebgewonnene Spielereien von James-Bond-Filmen ist der MINI aber kein gewöhnliches Auto, sondern ein wahres Wunderwerk der Technik, das mit einer Armbanduhr ferngesteuert werden kann.

In *The Bourne Identity* von 2002 ist Jason Bourne, gespielt von Matt Damon, ein ehemaliger Auftragsmörder der CIA, der sein Gedächtnis verloren hat und nun permanent in Todesgefahr schwebt. Sein Pendant ist die sympathische, abenteuerlustige Weltenbummlerin Maria, eigentlich Franka Potente, die ihr gesamtes, chaotisches Leben in einen alten, arg ramponierten Mini gepackt hat, der selbst an längst vergangene Hippie-Zeiten erinnert. Als der Verfolgte Maria um eine Mitfahrgelegenheit nach Paris bittet, willigt sie ein und steuert ihn nervenstark durch alle Krisen. Zu den besonders spannenden Momenten im Film zählt eine Verfolgungsfahrt durch Paris, bei der der kleine Wagen sämtliche Polizeikräfte abhängen kann – ein MINI macht eben alles möglich.

from his car. When he wants to carry a television set on the roof, a TV chair, or even a 4 m-high Christmas tree, the small car turns into a large van on the spot. The scene where the Mini Cooper is run over by a tank, unnoticed by its owner who is standing right alongside, is unforgettable.

The *Austin Powers* trilogy is a post-modern parody of British spy films of the 1960s. In the third part, *Austin Powers in Goldmember*, the new MINI graced the big screen in 2002 for the first time. Screenwriter and leading actor Mike Myers had lined up an extraordinary array of Hollywood stars, as well as six MINI Coopers in Union Flag kit, for this cinematic spectacular. In an allusion to the gimmicks beloved of the James Bond movies, here the MINI is no ordinary car but rather a true technological miracle that can be driven by remote control using a wristwatch.

In *The Bourne Identity* of 2002, Jason Bourne, played by Matt Damon, is a former CIA contract killer who has lost his memory and now lives permanently in mortal danger. His counterpart is the attractive, adventurous traveller Maria (Franka Potente), who has packed her whole chaotic life into a badly battered old Mini that looks like an escapee from the hippy era. When Bourne asks Maria for a lift to Paris, she agrees and drives him from crisis to crisis with nerves of steel. One of the film's most exciting scenes is a chase through Paris in which the small car manages to shake off the police force. A MINI really can make anything possible.

Ein Exponat der besonderen Art: In der zweiteiligen Installation *MINI KAPOOOW!* durchbricht ein MINI Paceman im Maßstab 1:1 räumliche Grenzen und erlebt eine Transformation der Materialien und Formen. Der Grundgedanke dieses Raumkonzepts beschreibt den MINI als ein sehr emotionales, Normen durchbrechendes Produkt mit ausgeprägter Persönlichkeit. Sportlich temperamentvoll setzt der Wagen zum Sprung in ein neues Universum an, in dem sich Farben und Materialien verwandeln und bisher nicht gekannte Erlebnisräume schaffen.

Die erste Phase zeigt das Heck des MINI Paceman als ein verchromtes, reales Fahrzeug, eine plastisch modellierte Skulptur im Raum. Erst zur Mitte hin zeichnet sich eine Metamorphose ab, bei der die nun folgenden Teile des Fahrzeugs auseinanderzufliegen scheinen, wobei die davon ausgehenden bunten Fäden diesen mit Energie aufgeladenen Prozess symbolisch untermalen.

In der zweiten Phase durchbricht der MINI deutlich sichtbar eine Mauer. In der sich nun eröffnenden Dimension löst sich das Fahrzeug von seiner bisherigen Form und lässt seine Front zu einem Ideenwerk aus Papier werden. Nur die Konturen sind noch zu erahnen. Durch die Vielzahl an Papier und Farbplättchen gewinnt der MINI Paceman eine neue Materialität und Gegenständlichkeit. Zugleich ist hiermit die Phase des Prototyping im Kreativprozess versinnbildlicht.

Schöpfer dieser Installation war das MINI Designteam um den Chefdesigner Anders Warming, das sich von den Ausdrucksmöglichkeiten des klassischen Comics inspirieren ließ. Präsentiert wurde *MINI KAPOOOW!* – auch diese Wortschöpfung hat ihre Wurzeln in der Comic-Kultur – bei der 52. Designmesse Salone del Mobile in Mailand 2013. Der Ideenreichtum war von Erfolg gekrönt: *MINI KAPOOOW!* wurde mit dem iF Gold Communication Award 2013 ausgezeichnet.

/

It was a very special kind of exhibit. In the two-part installation *MINI KAPOOOW!*, a full-size MINI Paceman breaks through from one space to another and is transformed in both materials and form. The basic idea of this spatial installation conceives the MINI in highly emotional terms, transcending norms with its distinctive personality. With its sporty spirit, the car leaps into a new universe in which colours and materials metamorphose to create hitherto unknown experiences.

In the first part, the rear end of the MINI Paceman is a real chrome-plated car, a three-dimensional sculpture in space. The metamorphosis only starts at the car's mid-point when subsequent parts of the vehicle seem to fly apart with colourful strings emerging from the back to underscore this energy-filled process symbolically.

In the second part, the MINI can be seen smashing through a wall. As this new dimension appears, the car is liberated from its previous form and its front end becomes a mass of ideas on paper, with only the silhouette still visible. Amid all the paper and coloured tabs the MINI Paceman takes on a new graphic and material identity – and the prototyping phase in the process of designing a car is thus brought to life.

The creators of this installation were the MINI design team led by Head of Design Andreas Warming, who had been inspired by the expressive possibilities of the classic comic strip. *MINI KAPOOOW!* – the word itself is rooted in comic-strip culture – was presented in 2013 at the 52nd design show Salone del Mobile in Milan. The team's inventiveness achieved great success, as *MINI KAPOOOW!* was honoured in 2013 with a Gold iF Communication Award.

John Lennon in Georg Harrisons Mini /
John Lennon in George Harrison's Mini

Bei einem Fahrzeug wie dem MINI, der früh Kultstatus gewann und seit bald sechzig Jahren im Rampenlicht steht, liegt es nahe, dass sich auch Künstler mit ihm befasst haben bzw. der prominente Kleinwagen zur »Leinwand« ästhetischer Betrachtung wurde. Den Anfang machte George Harrison, der seinen zunächst schwarz lackierten Mini, einen Austin Cooper S, Anfang 1967 neu gestalten ließ. Als Grundton wählte er ein warmes Rot, auf dem – inspiriert durch ein Buch über Tantrum Art – psychedelische Bilder und Symbole aufgetragen wurden. Bei den Dreharbeiten zum Beatles-Film *Magical Mystery Tour* im selben Jahr wurde der Wagen dann kurzerhand mit eingebunden und damit unsterblich.

Vorübergehend ging dieser Mini in den Besitz von Eric Clapton über, ehe er wieder an Harrison zurückgegeben wurde. Anlässlich des fünfzigsten Geburtstags der Marke wurde 2009 ein neuer MINI auf ähnliche Weise dekoriert und Olivia Harrison, der Witwe des 2001 verstorbenen Gitarristen, übergeben. Damit gelangte der Wagen in den Bestand der von George Harrison gegründeten Material World Charitable Foundation. Sichtlich gerührt, bemerkte Mrs Harrison: »George war ein begeisterter Mini-Fan und hätte sich bestimmt gefreut, diese neue Version mitzugestalten. Dass die Geburtstagsfeier des Mini auch unserer Stiftung zugute kommt, verleiht unserer erfreulichen Zusammenarbeit besonderes Gewicht.«

Ebenso vom Zeitgeist und Kunstgeschmack geprägt, ist das Ende der Siebzigerjahre entstandene MINI-Gemälde des französischen Künstlers Christian Claerebout. Vorwiegend bekannt für Darstellungen von Elefanten, für Reflektionen über die Finanzwelt und Kompositionen zur klassischen Musik, setzt sich Claerebout seit 1976 immer wieder mit dem MINI auseinander. Frei von aller zeitlichen Einordnung geht es ihm um das Erfassen des Charakters und der Ausstrahlung wie auch der Gestalt des Fahrzeugs. In dem Gemälde aus den Achtzigerjahren zeigt sich der Mini in beispielloser Einfachheit im klassischen Seitenprofil. Dabei verzichtete der Künstler auf die Ausarbeitung von Details und stellt die für MINI typische Kontur in den Vordergrund. Die Fläche wird betont, nicht der Bezug zum Räum-

When a car such as the MINI gains cult status early on and has been in the limelight for almost 60 years, it makes sense that artists too should turn their attention to it, and that the famous small car should be considered a 'canvas' for art. The trend was launched by George Harrison, who had his originally black Mini, an Austin Cooper S, redesigned early in 1967. He chose a warm red as the background colour over which psychedelic images and symbols, inspired by a book on Tantric art, were applied. The same year the car became a feature on the set of the Beatles' film *Magical Mystery Tour*, and thus attained immortality.

The car came briefly into the possession of Eric Clapton before it was returned to Harrison once more. On the brand's 50th birthday in 2009, a new MINI was decorated in a similar style and presented to Olivia Harrison, widow of the guitarist who had died in 2001. Thus the car came into the collection of the Material World charitable foundation that had been set up by George Harrison. Visibly moved Mrs Harrison commented: 'George was a huge Mini fan and he would have enjoyed creating this new version. The fact that MINI's anniversary celebrations will also benefit our Foundation has made it an enjoyable as well as meaningful collaboration'.

Equally imbued with the spirit and artistic taste of its time are the Mini paintings, created in the late 1970s by the French artist Christian Claerebout. Mainly known for his paintings of elephants, and works inspired by the world of finance or classical music, Claerebout has engaged with the MINI time and again since 1976. Unconcerned by any chronological classification, Claerebout aims to capture the character and charisma as well as the shape of the car. In a work from the 1980s the Mini is depicted with unparalleled simplicity in a classic side view. The artist did not worry about the depiction of details, foregrounding instead the typical MINI silhouette with an emphasis on the two-dimensional surface, rather than a representation of three-dimensional space. Claerebout's use of colour shows great flair, as the basic colours of red, blue and yellow form restless surfaces and nervous lines across the picture, contrasting with the dark grey of the back-

Mini von / by Paul Smith, 1999

lichen. Mit großem Elan trug Claerebout hier die Farbe auf. Die Grundfarben Rot, Blau und Gelb verteilen sich in unruhigen Flächen und nervösen Linien über dem Bild und bilden einen Kontrast zum dunklen Grau des Hintergrunds. Der spontane, kühne bis flüchtige Pinselduktus und die plakative Leuchtkraft der Farben erzeugen eine besondere Dramatik, als wollten sie einen Augenblick festhalten, und geben sich als Stilmittel der Pop-Art zu erkennen.

1999, anlässlich des 40. Geburtstags, wurde der klassische Mini von dem berühmten englischen Modedesigner Paul Smith gestaltet. Der »Meister der Streifen«, wie er auch genannt wird, ist ein bekennender Verehrer des MINI und sieht in ihm einen zeitlosen Klassiker. Bei der Gestaltung ließ er sich von den Streifen seiner Frühjahrs- und Sommerkollektionen inspirieren. Mit 86 Streifen in 26 verschiedenen Farben wurde der Wagen neu lackiert – noch heute fasziniert die Farbkombination.

Im selben Jahr wurde auch der Rocksänger David Bowie gebeten, einen klassichen Mini künstlerisch zu veredeln. Statt Farben, Formen oder Folien aufzutragen, ließ Bowie den Wagen komplett verchromen. Auch die Fenster wurden vollständig verspiegelt. Angesprochen auf die unkonventionelle Gestaltung, die den Wagen mit einer glänzenden, zweiten Haut überzog, bemerkte der Künstler, sie sei als ein Design gedacht, das sich nicht von seiner Umgebung unterscheiden will, sondern diese wiedergibt. So soll der MINI hier kein Objekt der Betrachtung sein, sondern das Umfeld des Betrachters, die Wirklichkeit spiegeln.

ground. The spontaneous brushstrokes, sometimes bold sometimes fleeting, and the bold luminosity of the colours produce a special kind of drama, as if they wanted to capture the moment – a clear stylistic nod to Pop art.

In 1999, on its 40th birthday, the classic Mini was customized by renowned English fashion designer Paul Smith. Also known as the 'master of stripes', Smith is an acknowledged admirer of the MINI which he considers to be a timeless classic. For his creation, he took inspiration from the stripes in his spring and summer collections. The vehicle was newly painted with 86 stripes in 26 different colours – and even today, this combination of colours continues to fascinate.

In the same year rock star David Bowie was invited to employ his artistic talents in customizing a classic Mini. Instead of applying colours, shapes or lamination, Bowie had the car entirely chromeplated. Even the windows were completely metallized. When asked about the unorthodox design that covered the vehicle with a shining second skin, the artist said that it was intended to be a design that could not be distinguished from its environment, but rather would reflect it. Thus the Mini should not be an object of contemplation, but one that reflected the viewer's own surroundings and reality.

Verchromter Mini von / Chrome-plated Mini by David Bowie, 1999

Christian Claerebout, *Mini*, ca. 1979

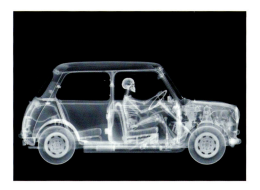

Nick Veasey, *Mini X-Ray*, 2012

Einen besonderen Bezug zu MINI hat auch der Brite Nick Veasey: »Durch den Mini genoss ich ein hohes Maß an Freiheit. Er prägte ganz eindeutig die Mobilität in der Nachkriegszeit, weil er erschwinglich war. Einen Mini konnten sich auch junge Leute leisten. Er war ausgesprochen schick und sehr leicht zu handhaben.« Durch Zufall war der Künstler mit der Technologie der Radiografie in Verbindung gekommen. Die besondere Ästhetik, die von Röntgenbildern ausgeht, brachte ihn im Lauf der Jahre mehrfach dazu, Gegenstände dem Röntgenblick auszusetzen und ihre innere Struktur zu enthüllen. Für den MINI als Bildgegenstand sprach seine Einfachheit in Bezug auf Technik und Design und sein Aufstieg zum Designklassiker. Für das Werk, das 2012 entstand, nahm Veasey einen klassischen Mini komplett auseinander und setzte jedes Bauteil den Röntgenstrahlen aus, anschließend baute er den Wagen am Computer wieder zusammen.

Mit seiner Röntgenfotografie schlägt Nick Veasey eine Brücke zwischen Kunst und Wissenschaft. Mit Hilfe einer bestimmten Technologie versucht er, die tatsächliche Beschaffenheit von Dingen, die wir glauben zu kennen, sichtbar zu machen. Denn unsere Welt, so der Künstler, ist mit Bildern geplagt. Röntgenstrahlen durchdringen die vordergründige Schönheit oder vertraute Oberfläche und führen zur wesensimmanenten, inneren Schönheit.

Unmittelbar an der Oberfläche eines klassischen Mini arbeitete der Brite Damien Hirst. Sein Werk aus dem Jahr 2000 trägt den Namen *Untitled* und zeigt 1400 Farbpunkte, die in einem regelmäßigen Muster über die Karosserie verteilt sind. Damit reiht sich der Mini ein in die durch Hirst 1986 begründete Werkgruppe der *Spot Paintings*.

Auch der neue MINI ab 2001 ist häufig Gegenstand ästhetischer Betrachtung. Im Prinzip steht hierbei weniger eine rein künstlerische Herangehensweise im Vordergrund, als die Umsetzung kreativer Ideen. Das Innovationspotenzial des MINI, das auf seinen Schöpfer Sir Alec Issigonis zurückgeht, soll auch heute Ausdruck finden. Eine Plattform für unkonventionelles Dekor und Design bietet die Marke seit Jahren beim Life Ball in Wien, einer gesellschaftlich viel beachteten Benefizveranstaltung, die Signale im Kampf gegen HIV und Aids setzen

British photographer Nick Veasey also has a special relationship with the MINI: 'I'd been given a great deal of freedom by them. They were undoubtedly a mainstay of affordable post-war mobility. Young people could afford them and they looked brilliant and handled superbly'. The artist discovered the technology of radiography by sheer accident. The special aesthetic produced by X-ray images led him over the years to subject objects repeatedly to the X-ray process to reveal their internal structure. The MINI's suitability as a subject was suggested by its simplicity in both technology and design, and its status as a design classic. In his 2012 work, Veasey completely stripped down a classic Mini and X-rayed each component, then re-assembled the vehicle on the computer.

With his X-ray photography, Veasey creates a bridge between art and science. By using this particular technology, he tries to reveal the actual nature of things we believe we already know. According to Veasey our world is plagued with images. X-rays penetrate their superficial beauty or familiar surface and lead us to their inherent internal beauty.

For British artist Damien Hirst it was the surface of a classic Mini that interested him. His 2000 work *Untitled* features 1,400 coloured spots distributed over the car's body in a regular pattern. This Mini forms part of Hirst's *Spot Paintings* series, which he began in 1986.

The new post-2001 MINI has also been the frequent subject of artistic endeavours. Such cases demonstrate more the application of creative ideas than a purely artistic approach. Even today the MINI's potential for innovation, that can be traced back to its creator Sir Alec Issigonis, cannot remain hidden. The brand has offered a platform for unorthodox décor and design for many years at the Life Ball in Vienna, a high-profile fund-raising event that supports the fight against HIV/AIDS. Well-known designers such as Donatella Versace, Mario Testino or Diane von Furstenberg sponsor a MINI that they customize individually in order to be auctioned.

Urban artist Andreas Preis forged new directions in art when he discovered the possibilities of using skateboards and snowboards as

Installation beim Mailänder / Installation at the Milan
Salone del Mobile, 2014

will. Namhafte Designer übernehmen die Patenschaft für einen MINI, den sie individuell gestalten und anschließend versteigern lassen. So tragen einige Kreationen die Handschrift von Donatella Versace, Mario Testino oder Diane von Furstenberg.

Neue Wege künstlerischer Gestaltung ging der Urban Art Illustrator Andreas Preis, der Skate- und Snowboards als Bildträger für sich entdeckt hat. Zum Saisonauftakt der Burton European Open im schweizerischen Laax gestaltete er nicht nur einen MINI Paceman, sondern entwickelte auch die sogenannten MINI Snow Beasts. Auf insgesamt fünf Snowboards erscheinen verschiedene Wappentiere, die symbolisch für den Schnee und die Bergwelt stehen. Der Steinbock repräsentiert die Ambition, der Adler die Vision, der Luchs die Anmut, der Wolf Teamwork und die Eule Weisheit. Mit sicherer Linienführung und Gespür für eine symbolhaltige Ornamentik und eindringliche Formensprache schuf Preis eine Kollektion außergewöhnlicher Darstellungen, welche ein herkömmliches Snowboard in ein neues, künstlerisch beachtenswertes Bildformat überführt.

Ein interaktives Licht- und Klangwerk der besonderen Art bot der Salone del Mobile 2014 in Mailand: In Zusammenarbeit mit dem Londoner Kunst- und Designkollektiv United Visual Artists (UVA) präsentierte MINI die Installation *Parallels*. Inspiriert von der Technologie des MINI Connected und ihren unbegrenzten Möglichkeiten der Vernetzung, schuf UVA eine mehrteilige Inszenierung, welche die Beziehung zwischen Mensch und Technologie neu beleuchtete. *Parallels* übersetzte den heute schon vielfach praktizierten grenzenlosen Übergang von Wohnung, Büro und Fahrzeug in eine erlebbare Kunstform. Dabei rückte das prägnanteste Element des neuen Interface-Designs von MINI in den Mittelpunkt: das Licht im LED-Ring, der das Center-Instrument rahmt und durch unterschiedliche Lichtimpulse mit dem Fahrer geräuschlos kommuniziert. Drei Ringe mit dem Durchmesser von jeweils 1,5 Metern projizieren Lichttunnel in den dunklen Raum. Der Besucher kann durch diese hindurchwandern und dank der Sensorik von Bewegungsmeldern mit ihnen in einen interaktiven Dialog treten.

his canvas. For the start of the Burton European Open in Laax, Switzerland he designed not only a MINI Paceman, but also the so-called MINI Snow Beasts. On a series of five snowboards he has painted different heraldic animals that symbolize the snow and the mountains. The ibex represents ambition, the eagle vision, the lynx gracefulness, the wolf teamwork and the owl wisdom. Using bold outlines, a feeling for symbolic ornamentation and a striking formal language, Preis has created an extraordinary collection that transforms a conventional snowboard into a new and remarkable work of art.

An unusual kind of interactive light and sound work appeared at the 2104 Salone del Mobile in Milan when MINI unveiled the installation *Parallels* in collaboration with the London art and design collective United Visual Artists (UVA). Inspired by the technology of MINI Connected and its unlimited networking possibilities UVA created a multi-part installation that shed new light on the relationship between humans and technology. *Parallels* translated the multiple everyday transitions from home to office to car into an artistic experience. The work focused on the most striking element of the MINI's new interface design – an LED laser ring that frames the centre instrument and communicates silently with the driver using a variety of light impulses. Three 1.5 m diameter rings project laser tubes into a darkened room. The visitor moves through the room, entering into an interactive dialogue using the sensor technology of motion detectors.

DESIGN ICONS

Der klassische Mini von 1959 ist rückblickend purer Ausdruck der Lebensfreude der Swinging Sixties. Designhistorisch wie auch in der verhaltenen Farbpalette folgte er jedoch der Ästhetik der Fünfzigerjahre. Ein hohes Maß an Funktionalität dominiert die Gestaltung, verbunden mit Reduktion und schlichter Eleganz.

Das nach 1960 vorherrschende Design richtete sich zunächst an Technik und modernen Produktionsmethoden aus, wich aber zunehmend einer emotionalen Komponente. Auch die Farbpalette ging mit der Zeit und wurde, wie auch Produkte anderer Sparten, bunter. Maßgeblichen Anteil an diesem Wandel hatte die »Plastic Revolution«: Verschiedene Kunststoffe wurden erfunden, die leicht, beliebig formbar und in allen Farben produziert werden konnten. Zudem gewann damals der Systemgedanke an Raum, eine Idee, die das Kombinieren eines Gegenstands aus Basiseinheiten vorsah und sich auch in Erfindungen wie dem Stapelstuhl oder dem Stapelgeschirr äußerte. Die Ergonomie wurde zur Leitidee, also die Handhabbarkeit eines Gegenstands bezogen auf die Bedürfnisse und Bewegungsabläufe des menschlichen Körpers. Sie sollte der gemeinsame Nenner aller Innovationen sein. Neben Formvielfalt, Farbigkeit und dem funktionalistischen Ideal der Moderne war das Design der Sechzigerjahre ebenfalls geprägt von unkonventionellen Lösungen, einem hohen Maß an Experimentierfreude, Emotionalität sowie dem Streben nach Lebensqualität. Hier fand der scheinbar grenzenlose Optimismus, das Vertrauen in die eigene Gestaltungskraft, durchmischt von Weltraumfantasien und Kunststoffträumen, seinen stärksten Ausdruck.

Hatte der klassische Mini gerade in seiner Schlichtheit eine ideale Projektionsfläche geboten und dabei vielfältig Akzeptanz gefunden, so erfüllt der MINI des beginnenden 21. Jahrhunderts andere Ansprüche: Er vermittelt seinem Besitzer stets das Gefühl, richtig angezogen zu sein. Sein Design ist weder futuristisch noch revolutionär, sondern modern, weltoffen und authentisch – in dieser Ausprägung wird der MINI den Weg in die Zukunft einschlagen. Dabei wird es faszinierend sein, zu beobachten, welche Parallelen sich zeitgleich im Produktdesign ergeben werden und welchen Einfluss das MINI-Design nicht nur in der Automobilbranche, sondern darüber hinaus ausüben wird.

/

In retrospect the classic Mini of 1959 is a pure expression of the *joie de vivre* characteristic of the Swinging Sixties. In terms of design, however, as with its restrained colour palette, it espoused the aesthetics of the 1950s. A high degree of functionality dominates the design, alongside compactness and simple elegance.

After 1960 the predominant design was initially oriented towards technology and modern production methods, although this increasingly gave way to more emotional factors. The old colour palette disappeared too over time and became more vibrant, like products in other sectors. The so-called 'Plastic Revolution' played a major part in this transformation. Different plastics that were light, easy to mould and that could be produced in all colours were being developed. In addition, this period saw the rise of modular design, whereby an object could be assembled from basic units, such as the stacking chair or stackable crockery. Ergonomics became the guiding philosophy, i.e. the usefulness of an object in relation to the needs and movements of the human body. It was planned as the common denominator in all innovations. Alongside variety of form, colourfulness and the functional ideal of modernity, 1960s' design was equally characterized by unorthodox solutions, a spirit of adventure in experimentation, emotionality, and the pursuit of quality of life. Seemingly boundless optimism and the confidence in one's own creative power, mixed with space-age fantasies and dreams of synthetic worlds found their most powerful expression during this era.

If the simplicity of the classic Mini was the ideal blank canvas that contributed to its widespread popularity, the MINI of the early 21st century fulfils other criteria. The MINI always makes its owner feel correctly dressed. Its design is neither futuristic nor revolutionary, but modern, open to the world and authentic – and it is with these attributes that the MINI will head off into the future. It will be fascinating to observe the influence that MINI design will exercise both in the automobile industry and beyond.

Anders als erhofft, konnte der Mini in den ersten Jahren, den frühen Sechzigern, nicht die Zielgruppen überzeugen, für die er geplant war: Trotz seines günstigen Preises war er für junge Käufer noch zu teuer, für besser Verdienende zu spartanisch und zu wenig repräsentativ. Stattdessen avancierte er schon bald zum Kultauto einer in London aufkommenden Bewegung, die als die Swinging Sixties bekannt geworden ist.

Ein Freund von Sir Alec Issigonis, der Fotograf Anthony Armstrong-Jones, der von der Queen in den Adelsstand erhoben wurde und besser bekannt ist unter dem Namen Lord Snowdon, war einer der ersten Besitzer. Begeistert vom Gokart-Feeling des kleinen Flitzers fuhr er mit dem klassischen Mini durch die Straßen Londons. Als Ehemann von Prinzessin Margaret hatte er die denkbar besten Verbindungen zum britischen Königshaus, und so fiel es ihm leicht, ein Treffen zwischen seiner Schwägerin Queen Elisabeth II. und dem MINI-Erfinder zu arrangieren. Nach einer kurzen Vorstellung des Fahrzeugs nahm die Königin neben Issigonis im klassischen Mini Platz und ließ sich eine Runde durch den Park von Schloss Windsor fahren – eine zunächst unscheinbare Begebenheit, die aber ausreichte, den Mini zum Gespräch der High Society zu machen.

Schon bald fand der populäre Schauspieler Peter Sellers Gefallen am klassischen Mini. Den ursprünglich spartanisch ausgestatteten Kleinwagen ließ er im Interieur veredeln und außen mit einem Überzug versehen, der an ein Korbgeflecht erinnerte. Dieses sogenannte Wickerwork-Trimm gefiel auch Fürst Rainier von Monaco so sehr, dass er sich eine ähnliche Luxusvariante des Mini bauen ließ. Peter Sellers indes war von dem Wagen mehr als begeistert. Er kaufte mindestens ein Dutzend Modelle, ließ diese mit erlesenem, mitunter auch ausgefallenem Geschmack individuell umgestalten, um sie anschließend zunächst seiner Ehefrau Britt Ekland, später auch seinen wechselnden Lebensgefährtinnen zu schenken.

Plötzlich war der Mini das begehrte Fahrzeug der Londoner Szene und gewann über alle gesellschaftlichen Schichten und Altersgruppen hinweg an Beliebtheit. Sein innovativer und nonkonformistischer

DIE SWINGING SIXTIES
/
THE SWINGING SIXTIES

In its first few years, during the early 1960s, the Mini did not win over the target groups it was aimed at, contrary to expectations. Despite its attractive price, the car was still too expensive for young buyers, and too spartan and lacking in prestige for the well-to-do. Instead, however, it soon turned into the cult car of the London phenomenon that became known as the Swinging Sixties.

One of the first Mini owners was photographer Anthony Armstrong-Jones, a friend of Sir Alec Issigonis, who was ennobled by the Queen and is better known as Lord Snowdon. Inspired by the go-kart feeling of the small runabout, Snowdon drove his classic Mini round the streets of London. As the husband of Princess Margaret Snowdon was well connected to the British royal family and so it was easy for him to arrange a meeting between his sister-in-law, Queen Elizabeth II, and the inventor of the Mini. After a short introduction to the car, the Queen sat next to Issigonis in a classic Mini and was driven through the grounds of Windsor Castle. What seemed at first sight like an insignificant event was enough to make the Mini a hot topic in high-society conversation.

Soon afterwards, the popular actor Peter Sellers fell in love with the classic Mini. He had the original spartan interior of the small car upgraded and on the exterior he had the panels re-painted with a basket motif. This so-called wickerwork trim also appealed to Prince Rainier of Monaco and he too had a similar luxury edition of the Mini made. Meanwhile, Peter Sellers' passion for the car continued. He bought at least a dozen models and had them individually customized in luxurious and at times unusual styles, giving them to his wife Britt Ekland and later to his variety of ladyfriends.

Suddenly, the Mini had become the must-have car on the London scene, popular among all social classes and age groups. Its innovative and non-conformist character chimed with the spirit of the 1960s, an era of increasingly progressive ideas, positive upbeat mood and a certain disrespect of convention.

It was as if the Mini had been born for this exciting period. Its legendary fame, iconic charisma and nostalgic charm all stem from the

1960s. It was 'like nothing else: it was a box, a simple container, into which people could squeeze anything and anybody they needed to [...] The Mini became the suitcase into which they packed their hopes', for the big city. 'In its otherness, the Mini was the symbol and anti-symbol of its time: it was cheeky, impudent, inquisitive, bold, rakish, dashing and brazen.'

A band from Liverpool, who had performed for the first time in August 1960 in a club in Hamburg under the name The Beatles, stood unchallenged at the heart of the soon-to-be-legendary Swinging Sixties. In 1963 their song *She Loves You* climbed the hit parade. The mop-top lads became the icons of their time, embodying for the youth of England and Western Europe successful self-liberation and rebellion against their parents' generation, and against a barren, uptight and repressive post-war atmosphere.

All You Need is Love became the critique of an unshakeably hierarchical society. If not a classless society it offered at least a classless experience, a way of living together that transcended barriers. In 1965, with *Satisfaction* and *My Generation*, The Rolling Stones and The Who created a similarly revolutionary force.

We know that the Beatles were also huge fans of the Mini. Ringo Starr had his Mini expanded to accommodate his drum-kit. Paul McCartney too owned a classic Mini, as did John Lennon who cruised round the City of London in it with George Harrison. Cliff Richard, Mick Jagger and members of The Who all owned Minis – no wonder, then, that such appearances increased sales of the car considerably.

The Swinging Sixties in London embodied an exceptional cultural period, similar to turn-of-the-century Vienna or 1920s Berlin. Demonstrations under the banner of 'Make Love not War' and calls for the abolition of nuclear weapons characterized the Swinging Sixties as much as drug consumption, pot parties and the demand for sexual liberation.

It is thought that this movement was triggered by the fate of Great Britain itself. The rise of London as a cultural centre went hand-in-hand with the country's political decline. During the 1956 Suez crisis,

Charakter entsprach dem Zeitgeist der Sechzigerjahre mit seinem zunehmend fortschrittlichen Denken, seiner Aufbruchsstimmung und einer gewissen Respektlosigkeit gegenüber gängigen Konventionen.

Der Mini wurde sozusagen in diese schwungvolle Ära »hineingeboren«. Ihr verdankt er seinen legendären Ruhm, seine ikonenhafte Ausstrahlung, seinen nostalgischen Charme. Er war »fast eine Kiste, eine Box, ein Container, seine Gestalt ... ohne Referenz, ... eine Schublade, in der man verstauen kann, was und wer auch immer man ist ... die Schatulle für das kleine oder auch große Glück«, das die Stadt bereithielt. »In seinem Anderssein war der Mini das Zeichen und Antizeichen seiner Epoche: cheeky, also frech, vorwitzig, dreist, kess, flott, unverschämt.«

Im Mittelpunkt der später legendären Swinging Sixties stand unbestritten eine Band aus Liverpool, die im August 1960 unter dem Namen The Beatles in einem Hamburger Club erstmals aufgetreten war. 1963 führte ihr Song *She Loves You* die Hitparaden an. Die Burschen mit den Pilzkopf-Frisuren wurden zum Paradigma ihrer Zeit, verkörperten sie doch für Englands wie auch Westeuropas Jugend die erfolgreiche Selbstbefreiung und Rebellion gegen die Elterngeneration, gegen eine öde, spießige und repressiv empfundene Nachkriegsatmosphäre.

All You Need is Love wurde zum Ausdruck der Kritik an einer hierarchisch strukturierten, unverrückbaren Gesellschaft; es sollte – wenn schon nicht die Klassenlosigkeit – so doch wenigstens ein klassenübergreifendes Erlebnis geben, ein Grenzen sprengendes Miteinander im Augenblick. Vergleichbar revolutionäre Sprengkraft erzeugten die Rolling Stones mit *Satisfaction* von 1965 und The Who im selben Jahr mit *My Generation*.

Von den Beatles ist bekannt, dass auch sie Gefallen am Mini fanden. Um sein Schlagzeug transportieren zu können, ließ Ringo Starr den Kleinwagen ausbauen. Ebenso besaß Paul McCartney einen klassischen Mini, nicht anders John Lennon, der gemeinsam mit George Harrison durch Londons City cruiste. Cliff Richard, Mick Jagger und die Bandmitglieder von The Who – sie alle waren Besitzer eines Mini.

The Rolling Stones, 1963

Britain had learned painfully what it meant to be dependent on Middle Eastern oil imports, and to no longer be able to assert oneself with military might alone. Even though Great Britain numbered among the victorious Allies after World War II, the following 20 years saw the almost complete loss of its empire. Spy scandals and affairs among prominent politicians dominated the headlines, and the Northern Irish conflict was already casting its shadows. But in this same decade English cities became the internationally recognized hubs of new fashions and trends. At first, the movement was confined to a small artistic élite. Fine artists, designers, writers, fashion designers and film-makers embarked upon new experimental and provocative journeys. Their revolution became a recognized style, and their means of expression adopted colourful, lightweight, mobile, condensed, small formats. It was no coincidence that such minimalism found approval through the short skirt and the compact car.

Great Britain did not counter this trend with a rigid conservatism, but instead transformed itself into a highly modern country. For example, the death penalty was abolished, as was theatre censorship. Abortion, homosexuality and suicide were no longer considered crimes. The school system was reformed, and university entrance made easier. And London was always at the forefront. Here, in the pulsating heart of European Pop culture, an as yet unknown youth culture blossomed, along with the modern possibilities of mass communication. The Kinks' *Waterloo Sunset* is a legendary song dedicated to the London of those years, a romantic ballad that describes the sunset over Waterloo railway station on the Thames.

Kein Wunder, dass sich entsprechende Auftritte für den Wagen als besonders absatzfördernd erwiesen.

London und die Swinging Sixties stellen eine kulturelle Ausnahmeerscheinung dar, vergleichbar der Jahrhundertwende von 1900 in Wien oder dem Berlin der Zwanzigerjahre. Demonstrationen mit »Make Love not War«-Parolen und Forderungen nach Abschaffung von Kernwaffen bestimmen die schwungvollen Sechzigerjahre ebenso wie Drogenkonsum, Hasch-Parties und der Ruf nach sexueller Freizügigkeit.

Den Auslöser für diese Bewegung glaubt man im Schicksal Großbritanniens selbst zu finden. Der Aufstieg Londons zum kulturellen Mittelpunkt ging einher mit einem politischen Niedergang des Landes. Während der Suezkrise 1956 hatten die Briten schmerzhaft erfahren müssen, was es bedeutet, von Ölimporten aus dem Nahen Osten abhängig zu sein und sich mit militärischer Stärke allein nicht mehr behaupten zu können. Zählte Großbritannien nach dem Zweiten Weltkrieg zum Kreis der Siegermächte, ging sein Empire in den folgenden zwanzig Jahren doch fast völlig verloren. Spionageskandale und Affären der Politprominenz beherrschten die Schlagzeilen, der Nordirland-Konflikt warf seine Schatten voraus. Doch just in diesem Jahrzehnt wurde die englische Metropole zur weltweit beachteten Plattform neuer Moden und Strömungen. Zunächst blieb die Bewegung einer kleinen Elite von Kulturschaffenden vorbehalten. Bildende Künstler, Designer, Schriftsteller, Modemacher und Filmer gingen neue Wege, experimentierten und provozierten. Ihre Revolte wurde zum anerkannten Stil. Ihre Ausdrucksformen waren das Bunte, das Leichte, Mobile, Komprimiert-Kleinformatige. Nicht zufällig fand das Minimale seine Beliebtheit im sichtbaren Rock und Kleinwagen.

Großbritannien begegnete dieser Strömung nicht mit verhärtetem Konservatismus, sondern wandelte sich in einen Staat modernster Prägung: So wurde die Todesstrafe abgeschafft, ebenso die Theaterzensur. Abtreibung, Homosexualität und Selbstmord galten nicht mehr als Verbrechen. Das Schulwesen wurde reformiert, der Zugang zur Universität erleichtert. In erster Reihe fand sich stets London. Denn hier pul-

Peter Sellers und / and Britt Ekland, 1965

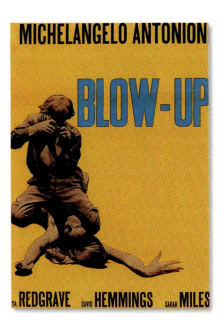

Filmplakat zu / Film poster for
Blow-Up, 1966

sierte das Herz der europäischen Pop-Kultur, hier blühte eine bisher nicht gekannte Jugendkultur auf, begleitet von den modernen Möglichkeiten der Massenkommunikation. Mit *Waterloo Sunset* widmete die Rockband The Kinks dem London dieser Jahre einen legendären Song, eine romantische Rockballade, die den Blick auf den Bahnhof Waterloo an der Themse bei Sonnenuntergang beschreibt.

Filmisch fanden die Swinging Sixties ihre treffendste Darstellung in Michelangelo Antonionis großem Klassiker *Blow-Up* aus dem Jahr 1966. Der Film spiegelt das Lebensgefühl einer neuen Generation wider, ihre Lebenslust, Sinnlichkeit und ihren Hang zur Apathie. Es ist die Geschichte eines Fotografen namens Thomas – gespielt von David Hemmings –, der mit einem Fotomodell alias Veruschka von Lehndorff arbeitet und sie verbal angeht, als handele es sich um einen Liebesakt. In einem Londoner Park fotografiert er heimlich ein Paar und dokumentiert beiläufig einen Mord. Der Film verwischt die Grenze zwischen Realität und Imagination, lässt Zweifel an der Wahrheit von Wirklichkeit und präsentiert mit Thomas, dem Fotografen, eine ratlos-rastlose, unsichere und unbestimmte Figur, die sich treiben lässt und für die das Triviale so viel Bedeutung hat wie das Spektakuläre. Unstet greift der junge Mann alles auf, bekundet Interesse, um es im nächsten Moment wieder fallen zu lassen.

Am engsten verbunden mit dem Phänomen der Swinging Sixties ist die Carnaby Street, die durch ihre unzähligen Musik- und Modegeschäfte bekannt wurde und in Westeuropa als Trendmeile galt. Wer »Youthquaker«, einfach hip oder gar Hippie sein oder einfach nur Drogen kaufen wollte, ging dort shoppen. In dieser Ära wurden die Grundmuster einer Jugendkultur gelegt, die in vielfältiger Weise bis heute – immer wieder variiert – lebendig ist.

Doch auch die schwungvollen Sechzigerjahre gingen vorüber. London war um eine Facette reicher und nun vor allem bei Besuchern aus aller Welt begehrt, die dem Glanz jener Zeit nachträumten. Der Mini, der hierbei eine bedeutende Rolle gespielt hatte, konnte noch über Jahre hinweg vom Mythos der Swinging Sixties profitieren – ein Erfolg, der lange Zeit die allmählich sinkenden Absatzzahlen übersehen ließ.

The most apposite image of the Swinging Sixties on the big screen appeared in Michelangelo Antonioni's classic 1966 film *Blow-Up*. The movie reflects the attitudes of a new generation, their enjoyment of life, their sensuality and their propensity to boredom. It tells the story of Thomas (played by David Hemmings), a photographer working with a model (Veruschka von Lehndorff), who talks to her during the shoot as if they are making love. In a London park he secretly photographs a couple and accidentally documents a murder. The film blurs the boundaries between reality and the imagination, questioning the truth of reality and depicting Thomas, the photographer, as a confused, restless, uncertain and indecisive character who goes with the flow, and for whom trivialities are just as important as the essentials. The young man behaves erratically, embracing and taking an interest in everything, only to drop it again the next moment.

At the heart of the phenomenon of the Swinging Sixties lies Carnaby Street, famous for its many music shops and boutiques, and well known in Western Europe as the street of fashion. For anyone who wanted to be part of the 'youthquake', be hip or even a hippy, or just wanted to buy drugs, this was the place to shop. During this period the fundamental characteristics of youth culture were laid down, and they live on today in all their manifold variety.

But even the Swinging Sixties had to come to an end. London had discovered a new side to its personality, and was now in demand by visitors from all over the world who still dreamed of that glamorous period. The Mini had played an important role and profited for many years from the legend of the Swinging Sixties – a success that enabled the gradual reduction in sales to be overlooked for a long time.

Das Modeleitbild der Sechzigerjahre ist unbestritten der Minirock. In einem einzelnen Bekleidungsstück gipfelt die Haltung einer ganzen Epoche, so scheint es. Wenn es stimmt, dass ein Modestil einer kulturellen Uhr gleicht, sich als Zeitmesser einer modernen Gesellschaft erweist, ist der Minirock in der Tat ein solches Zeichen. Er verkörperte Sex-Appeal, Dynamik, Freizügigkeit und wurde zum Ausdruck weiblicher Selbstbestimmung.

Frauen fast jeden Alters und aus allen Schichten trugen ihn, weil sie ihn als antihierarchisch und antielitär empfanden. Als Erfinderin des ultrakurzen Rocks gilt Mary Quant. Ihre Skizzen und Entwürfe von Hängekleidern in A-Linie sehen eine kniefreie Saumlinie vor und reichen bis in die Saison 1959/60 zurück – jene Jahre, in denen das gleichnamige Automobil der Weltöffentlichkeit präsentiert wurde.

Schon 1955 hatte Mary Quant in der Londoner Kings Road eine Boutique eröffnet, das Bazaar. Unzufrieden mit den bestehenden Kollektionen, begann sie, aus billigen Stoffen Mode nach eigenen Vorstellungen zu entwerfen. Chelsea, das Viertel, in dem sie sich niedergelassen hatte, war bis dato eher rückständig, doch wurde es um 1960 herum zum Treffpunkt von Künstlern, Schriftstellern und Theaterleuten.

The fashion icon of the 1960s is undoubtedly the miniskirt. The attitude of an entire era culminated, it seems, in a single item of clothing. If we can truly say that a fashion is like a cultural clock, the chronometer of a modern society, then the mini skirt is indeed such a sign. It embodied sex appeal, dynamism, permissiveness and became the expression of female self-determination.

Women of almost all ages and all classes wore it because they perceived it to be anti-hierarchical and anti-elitist. Mary Quant is the inventor of this ultra-short skirt. Her sketches and designs for unstructured, above-the-knee, A-line dresses emerged back in the 1959/60 season – the years when the car of the same name was first unveiled to the public.

Mary Quant had opened a boutique called Bazaar on London's Kings Road in 1955. Dissatisfied with the clothes available in the shops, she began to design her own collection in her own style, and made of cheap fabrics. The district of Chelsea, where Mary had set up shop, was somewhat old-fashioned at that time, but by 1960 but it had become the meeting place of artists, authors and theatre folk.

In the 1950s the dominant ideal was a feminine silhouette, one that emphasized the female body and was accompanied by perfect make-up and a carefully co-ordinated hairstyle. The new fashion, by contrast, could not have been more different. Mary Quant wanted nothing to do with haute couture's wasteful use of fabrics. The cut of her clothes was completely simplified, the silhouette was now straight and geometric, and the feminine body was re-defined. Body-skimming contours made waist and bust invisible and instead focus was directed to legs and knees by the considerably shorter hemlines. Knee-socks and tights – without which the miniskirt was inconceivable – completed the image. Nude-look tights became the symbol of femininity and fetishized the leg. Frequently, these short skirts and unstructured dresses recalled schoolgirls' gymslips, as the miniskirt accentuated a decidedly girlish appearance. With collections such as *Lolita*, *School Girl* and *Good Girl* Mary Quant brought her customers new freedoms and opportunities for experimentation.

Mary Quant in ihrem Mini / Mary Quant in her Mini

In den Fünfzigerjahren hatte das Ideal einer femininen Linienführung dominiert, das den weiblichen Körper betonte und perfektes Make-up sowie eine sorgsam abgestimmte Frisur forderte. Die neue Mode hingegen bot den völligen Kontrast: Mary Quant distanzierte sich vom verschwenderischen Umgang der Haute Couture mit Stoffen. Die Schnitte wurden stark vereinfacht, die Linienführung fiel nun gerade und geometrisch aus, der weibliche Körper wurde neu definiert: Taille und Brust wurden durch körperferne Linien aus dem Blick genommen, dafür das Interesse durch den deutlich gekürzten Rocksaum auf Beine und Knie gelenkt. Dazu passten Kniestrümpfe oder Strumpfhosen, ohne die der Minirock nicht denkbar gewesen wäre. Der transparente Strumpf war das Zeichen von Weiblichkeit und fetischisierte das Bein. Vielfach erinnerten die kurzen Röcke und Hängerkleidchen an Schulmädchen-Kittel. Der Minirock unterstrich ein betont mädchenhaftes Äußeres. Mit Kollektionen wie *Lolita*, *School Girl* oder *Good Girl* erlaubte Mary Quant ihren Kundinnen neue Freiheiten und Selbstversuche.

1962 wurde die britische *Vogue* auf Mary Quants Mode aufmerksam und verhalf ihr zum Durchbruch. Aus ihrer Boutique wurde ein gut florierendes Modeunternehmen, das Geschäfte auf der ganzen Welt belieferte. Neben dem Minirock trugen auch Strumpfhosen, Make-up und Accessoires ihr Label. 1966 erhielt Mary Quant für ihre Verdienste um den englischen Modeexport den Order of the British Empire.

Für ihre Models hatte Mary Quant von Vidal Sassoon, dem Friseur in der Bond Street, einen praktischen Kurzhaarschnitt kreieren lassen, der zum modernen Look wesentlich beitrug. Das Topmodel der Swinging Sixties war unbestritten Lesley Hornby, die 1965 mit 16 Jahren entdeckt worden war und ein Jahr später einen Vertrag beim britischen Modemagazin *Woman's Mirror* unterschrieb. Besser bekannt als »Twiggy« (deutsch = Zweiglein), verkörperte sie das knabenhaft-androgyne Körperideal der Zeit: eine Kindfrau mit kaum gerundeter Figur und langen, dünnen Gliedern. War ihre Figur auch weniger verführerisch, so faszinierten ihre großen Augen und ein traurig-unerfüllter, sehnsuchtsvoller Gesichtsausdruck.

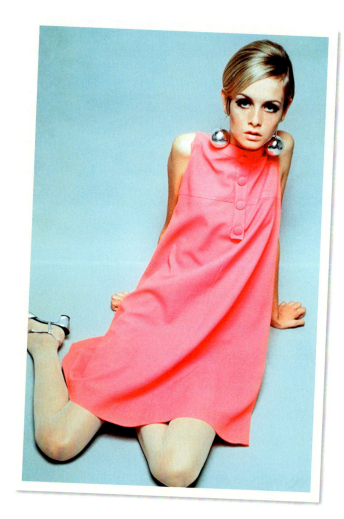

Twiggy, 1966

British *Vogue* magazine discovered Mary Quant in 1962 and helped her achieve her breakthrough. Her boutique became a highly successful fashion business that supplied shops across the world. Soon tights, make-up and accessories, as well as the miniskirt, bore the Mary Quant label. In 1966 Mary Quant received the OBE for services to English fashion exports.

Vidal Sassoon, a Bond Street hairdresser, was commissioned by Mary Quant to create a practical short haircut for her models that contributed much to the modern look. The top model in the Swinging Sixties was undoubtedly Lesley Hornby, who had been discovered in 1965 at the age of 16 and a year later signed a contract with British fashion magazine *Woman's Mirror*. Better known as Twiggy, Hornby embodied the boyish androgyny that was idealized at the time – a child-woman with hardly any curves and long, thin limbs. Although her figure might have been less than seductive, her large eyes and expression of sad, unfulfilled longing captured the public's imagination.

2

GESCHICHTE
/
HISTORY

MARKENGESCHICHTE / BRAND HISTORY

1959
- In August the British Motor Corporation (BMC) unveils a small four-seater badged as the Austin Seven and Morris Mini-Minor – the 'classic Mini' is born.
- After being lent a classic Mini by its designer Sir Alec Issigonis, John Cooper waxes enthusiastic about the car's racing qualities.

1960
- Presentation of the Mini Van, followed by the Austin Seven Countryman and Mini Traveller.

1961
- John Cooper and his association with Mini lead to a historic development: the first Mini Cooper appears on roads and racetracks.
- The Mini Pick-up is launched.
- Riley Elf and Wolseley Hornet are launched.

1963
- The classic Mini wins the Alpine Rally.
- Launch of the Mini Cooper S.

1964
- Paddy Hopkirk wins the Monte Carlo Rally in a classic Mini.
- The first off-road Mini: the Mini Moke.
- The classic Mini undergoes a major modification with the introduction of Hydrolastic suspension, developed by Alex Moulton (withdrawn again 1971).

1965
- The first classic Mini with automatic transmission enters the marketplace.
- Timo Mäkinen wins the Monte Carlo Rally in a classic Mini.
- Barely six years after its launch, Sir Alec Issigonis drives the 1-millionth classic Mini off the assembly line.

1967
- Mini triumphs for the third time in the Monte Carlo Rally, this time with 'Flying Finn' Rauno Aaltonen at the wheel.
- Sir Alec Issigonis is made a member of the Royal Society, Britain's most prestigious scientific institute.

1969
- In addition to a visually updated version of the classic Mini, the Clubman range is introduced.
- In autumn, British Leyland begins marketing the classic Mini, which BMC had previously sold under the Austin and Morris badges.
- The Mk II series and Cooper (except Mini Cooper S) versions are phased out, as are notchback saloons with the Riley and Wolseley badges.

1972
- The 3-millionth classic Mini leaves the Longbridge plant.

1984
- To mark the 25th anniversary, the Mini 25 special edition based on the Mini Mayfair is launched in a limited production run of 5,000.

1986
- The 5-millionth classic Mini rolls out of the factory.

1988
- Farewell to the creator of the classic Mini: Sir Alec Issigonis dies on 2 October 1988 aged 81.

1990
- The Mini Cooper goes back on sale.

1991
- The first classic Mini with a catalytic converter is launched.
- German dealer Lamm creates the first Mini convertible as a special model. Only 75 examples of the Lamm Cabriolet are built, but the idea lives on.

1992
- After a production period of more than 30 years, the 998 cc version of the classic Mini is phased out. From now on only the 1.3-litre model is available.

1993
- Inspired by the success of the Lamm Cabriolet, Rover launches its own convertible. Based on the Mini Cooper, it has luxury specifications and is the most expensive classic Mini model to date.

1994
- 35 years after the birth of the first classic Mini, Rover launches the Mini 35 special edition.
- 30 years after his unforgettable triumph, Paddy Hopkirk takes part in the Monte Carlo Rally, again in a classic Mini and once again as number 37.
- On 29 January 1994 BMW buys out the British company Rover.

1959
- Im August stellt die British Motor Corporation (BMC) einen kleinen Viersitzer unter dem Namen Austin Seven und Morris Mini-Minor vor.
- Der Erfinder des Mini, Sir Alec Issigonis, leiht dem Motorenentwickler John Cooper einen klassischen Mini und begeistert ihn für die Rennqualitäten des Autos.

1960
- Der Mini Van wird präsentiert, der Austin Seven Countryman und der Mini Traveller folgen.

1961
- John Cooper kooperiert mit Mini: Der erste Mini Cooper gibt sein Debüt auf Straßen und Rennstrecken.
- Der Mini Pick-up wird vorgestellt.
- Riley Elf und Wolseley Hornet kommen auf den Markt.

1963
- Der klassische Mini setzt sich gegen scheinbar unüberwindbare Rivalen durch. Er gewinnt die Alpine Rally.
- Der Mini Cooper S kommt neu ins Programm.

1964
- Paddy Hopkirk gewinnt in einem klassischen Mini die Rallye Monte Carlo.
- Der Mini Moke als erster Offroad-Mini geht in Produktion.
- Alex Moulton entwickelt für den Mini eine hydrolastische Aufhängung, die bis 1971 verbaut wird. Diese Art der Federung war bisher Limousinen vorbehalten.

1965
- Der erste klassische Mini mit Automatikgetriebe kommt auf den Markt.
- Timo Mäkinen gewinnt in einem klassischen Mini die Rallye Monte Carlo.
- Nur sechs Jahre nach Produktionsstart fährt Sir Alec Issigonis den einmillionsten klassischen Mini von der Montagestraße.

1967
- MINI triumphiert zum dritten Mal bei der Rallye Monte Carlo – diesmal mit dem »fliegenden Finnen« Rauno Aaltonen am Steuer.
- Mitglied der Royal Society: Sir Alec Issigonis wird in das renommierteste wissenschaftliche Institut Großbritanniens aufgenommen.

1969
- Die Clubman-Serie wird eingeführt. Sie zählt zu den optisch modernisierten Varianten im Programm.
- Ab Herbst vertreibt British Leyland den klassischen Mini. Zuvor wurde er von BMC unter den Marken Austin und Morris vermarktet.
- Mehrere Modelle werden eingestellt, darunter die Stufenheck-Limousinen unter den Markennamen Riley und Wolseley.

1972
- Der dreimillionste klassische Mini verlässt das Werk in Longbridge.

1984
- Zum 25-jährigen Jubiläum wird die Mini 25 Special Edition auf der Grundlage des Mini Mayfair aufgelegt. Es handelt sich hierbei um eine limitierte Serie von 5000 Fahrzeugen.

1986
- Der fünfmillionste klassische Mini verlässt das Werk.

1988
- Der Schöpfer des klassischen Mini, Sir Alec Issigonis, stirbt am 2. Oktober im Alter von 81 Jahren.

1990
- Der Mini Cooper kommt wieder auf den Markt.

1991
- Der erste klassische Mini mit Katalysator wird vorgestellt.
- Der deutsche Autohändler Lamm kreiert das erste Mini Cabrio als Sondermodell. Obwohl insgesamt nur 75 »Lamm-Cabriolets« gebaut werden, wird die Idee weitergetragen.

1992
- Die Produktion der 998cc-Version des klassischen Mini wird nach mehr als dreißig Jahren eingestellt. Ab diesem Jahr sind nur noch die 1,3-Liter-Modelle erhältlich.

1993
- Angeregt von dem Erfolg des Lamm-Cabriolets bringt Rover sein eigenes Cabrio auf den Markt. Die auf dem Mini Cooper basierende Luxusausführung ist das bisher teuerste klassische Mini-Modell.

1994
- 35 Jahre nach der Schöpfung des ersten klassischen Mini legt Rover die Mini 35 Sonderedition auf.
- Paddy Hopkirk nimmt dreißig Jahre nach seinem unvergesslichen Triumph erneut an der Rallye Monte Carlo teil – wiederum in einem klassischen Mini und zum zweiten Mal mit der Nummer 37.
- BMW übernimmt am 29. Januar die britische Firma Rover.

1996
- The 1,275 cc engine becomes the standard unit of the classic Mini, which now also features multi-point fuel injection.
- The Mini convertible is phased out. Total production amounted to 1,081 units.

1997
- 30 years after the last Mini victory in the Monte Carlo Rally, the Mini ACV 30 concept lines up at the start.
- Rover unveils the concept Spiritual for the new Mini at the Frankfurt Motor Show.

1998
- The classic Mini makes it into the Guinness Book of Records: with a total production run of 5.3 million, it is Britain's most popular car.
- Fashion designer Paul Smith creates his own Mini Limited Edition. It has special blue paintwork and is striking for its elegant simplicity.

2000
- The last of the classic Minis leave the factory: the Mini Seven, Mini Cooper and Mini Cooper Sport. For the German market, there is still the Mini Knightsbridge, but on 4 October 2000 the time has come to say goodbye. The last classic Mini bearing the number 5,387,862 rolls off the assembly line in Longbridge.

2001
- The new MINI has arrived. The MINI One and the MINI Cooper are unveiled to a delighted public. The classic Mini has become the so-called new MINI.

2002
- A sportier and more dynamic MINI Cooper S is let loose on the streets.

2004
- The 'always open' era begins with the MINI Convertible that drops its soft top for the world.
- The MINI XXL, a six-seater stretch version, is presented during the Olympic Games in Athens.
- It's off to the racetrack: the MINI CHALLENGE is held for the first time.

2006
- The new MINI is launched. A new generation of engines guarantees that pure go-kart feeling, while the interior has been completely re-designed.

2007
- MINI production reaches 1,000,000.
- The MINI Clubman is launched. It re-interprets the classic shooting brake concept in a modern way, highlighting sporty flair and functionality with its strikingly elongated roof-line and unique door arrangement.

2009
- MINI turns 50.
- MINI United – 25,000 enthusiasts from more than 40 countries converge on Silverstone to wish MINI a happy 50th birthday.
- MINI demonstrates its potential for further enhancement of driving pleasure and diversity through another spectacular concept car: the MINI Beachcomber Concept.

2010
- Launch of the MINI Countryman. For the first time, a production MINI can boast four doors and a large tailgate.

2011
- Driving fun in a new dimension: the MINI Rocketman Concept fuses traditional values with innovative technology to redefine the brand's core principle of the creative use of space.
- MINI expands its product family with the launch of the MINI Coupé.

2012
- MINI pushes ahead with the expansion of its model family. The latest addition to the range is the MINI Roadster, the sixth model in the brand's current line-up and the first open-top two-seater in its history.
- MINI presents the world's first Sports Activity Coupé in the premium small and compact vehicle segment; MINI Paceman is the seventh model in the brand family.

2013
- The new MINI makes its worldwide debut at Plant Oxford and pairs tradition with the go-kart feeling you know and love. The New MINI. The New Original.

2014
- Joan 'Nani' Roma wins the 2014 Dakar Rally in a MINI ALL4 Racing. This is the third victory for a MINI ALL4 Racing car from the X-raid team in four years since their debut in 2011.
- At the 2014 Geneva Motor Show MINI presents the MINI Clubman Concept, showcasing a new brand philosophy for a higher class of automobile.
- MINI Superleggera TM Vision – this exclusive interpretation of an open-top two-seater is created by MINI and Touring Superleggera. A unique model that blends the traditional with the modern.
- MINI expands its model range with the addition of an entirely new body variant – the MINI 5-door.

1996
- Der 1275cc-Motor wird serienmäßig im klassischen Mini verbaut, der ab jetzt mit einer Multi-Point-Kraftstoffeinspritzung ausgestattet ist. Insgesamt wurden 1081 Fahrzeuge hergestellt.
- Das Mini Cabrio wird nicht mehr produziert.

1997
- Dreißig Jahre nach dem letzten Sieg eines Mini bei der Rallye Monte Carlo erscheint der Mini ACV 30 am Start.
- Rover stellt das Konzept Spiritual für den neuen MINI auf der Frankfurter Automobilmesse vor.

1998
- Der klassische Mini schafft es in das Guinness-Buch der Rekorde: Mit einer Gesamtproduktionsstückzahl von 5,3 Millionen ist er Großbritanniens beliebtestes Auto.
- Modedesigner Paul Smith kreiert seine eigene Mini Limited Edition. Das Fahrzeug mit einer blauen Sonderlackierung besticht durch seine schlichte Eleganz.

2000
- Mit dem Mini Seven, dem Mini Cooper und dem Mini Cooper Sport verlassen die letzten klassischen Mini das Werk. Auf dem deutschen Markt ist zu diesem Zeitpunkt noch der Mini Knightsbridge erhältlich, bevor auch dessen Produktion am 4. Oktober 2000 eingestellt wird. Der letzte klassische Mini mit der Nummer 5387862 rollt in Longbridge von der Montagestraße.

2001
- Der neue MINI steht nun am Start. Der MINI One und der MINI Cooper werden einem begeisterten Publikum vorgestellt. Der klassische Mini hat sich zum sogenannten neuen MINI entwickelt.

2002
- Der MINI Cooper S, ein sportlicheres und dynamischeres Modell, ist ab jetzt im Straßenverkehr mit dabei.

2004
- Die Cabrio-Saison beginnt: Zur Freude der Öffentlichkeit lässt der MINI Cabrio sein Faltverdeck fallen.
- Der MINI XXL, eine Stretchlimousine mit sechs Sitzen, wird während der Olympischen Spiele in Athen präsentiert.
- Der MINI Challenge zeigt sich erstmals auf der Rennstrecke.

2006
- Der aktuelle MINI wird vorgestellt. Eine neue Motorengeneration sorgt für pures Gokart-Gefühl, der Innenraum zeigt sich in frischem Design.

2007
- Die MINI-Produktion erreicht die 1000000-Marke.
- Der MINI Clubman wird vorgestellt. Er interpretiert die klassische Shooting-Brake-Karosserie modern und verbindet sportliches Flair und Funktionalität. Auffällig an ihm sind die verlängerte Dachlinie und die einzigartige Türanordnung.

2009
- Der MINI wird fünfzig Jahre alt.
- MINI United: 25000 Fans aus mehr als vierzig Ländern trafen sich in Silverstone, um MINI zum fünfzigsten Geburtstag zu gratulieren.
- Sorgt für noch mehr Fahrspaß: Der MINI Beachcomber Concept als weiteres spektakuläres Konzeptfahrzeug.

2010
- Der MINI Countryman wird vorgestellt. Er ist der erste MINI mit vier Türen und einer großen Heckklappe.

2011
- Fahrspaß in einer neuen Dimension: Der MINI Rocketman Concept vereint traditionelle Werte mit innovativer Technologie. Damit wird das Grundprinzip der kreativen Raumnutzung der Marke neu definiert.
- MINI erweitert seine Produktfamilie mit der Einführung des MINI Coupé.

2012
- MINI führt die Erweiterung seiner Modellfamilie fort. Modell Nummer sechs ist der MINI Roadster, der erste offene Zweisitzer in der Geschichte der Marke.
- MINI präsentiert das weltweit erste Sports Activity Coupé in der gehobenen Klein- und Kompaktwagenklasse. Der MINI Paceman ist das siebente Modell der Markenfamilie.

2013
- Der neue MINI feiert sein weltweites Debüt im Werk in Oxford und vereint Tradition mit dem legendären Gokart-Gefühl.

2014
- Joan »Nani« Roma gewinnt die Rallye Dakar 2014 in einem MINI ALL4 Racing. Dies ist der dritte Sieg eines MINI ALL4 Racing aus dem X-raid-Team in vier Jahren seit der Gründung des Teams 2011.
- Symbol einer neuen Markenphilosophie: MINI präsentiert auf dem Genfer Autosalon 2014 das neue MINI Clubman Concept, das sich auf eine höhere Fahrzeugklasse ausrichtet.
- Vereint Tradition und Moderne: MINI und Touring Superleggera entwickeln gemeinsam einen neuen exklusiven und offenen Zweisitzer, den MINI Superleggera™ Vision.
- MINI erweitert sein Modellangebot mit einer neuen Karosserievariante – dem MINI Fünftürer.

Das Markenemblem von MINI hat eine wechselvolle Geschichte. Seine Wurzeln reichen zurück bis in die späten Vierzigerjahre. William Morris, der Gründer und Inhaber des britischen Fahrzeugherstellers Morris Motor Company, war wegen seiner Verdienste um die Automobilindustrie sowie seines karitativen Engagements geadelt worden und seit 1938 berechtigt, den Titel des Viscount Nuffield zu führen. Sein Wappen zierte ein roter Ochse, der sich über drei blauen Wellen – dem Symbol der Stadt Oxford – erhob und von einem beflügelten Rad gerahmt wurde. In Erinnerung an ihren Unternehmensgründer leitete die Firma Morris ihr Logo von diesem Wappen ab und schmückte damit unter anderem 1948 das erfolgreiche Modell des Morris Minor.

Das Motiv des geflügelten Rades folgt einer Tradition, die im Zusammenhang mit allegorischen Eisenbahndarstellungen steht und dessen Herkunft bis in die Mitte des 19. Jahrhunderts zurückreicht. Genauer betrachtet, handelt es sich um eine fiktive Kombination aus Rad und Flügeln, ein merkwürdiges Mixtum aus den Geschwindigkeits-Chiffren des Luft- und Landverkehrs. Der natürliche Vogelflug und das Rad von Menschenhand verbinden sich zu einer symbolisch aufgeladenen Figur, die höchstes Tempo verspricht. Das Emblem erfreute sich großer Beliebtheit und wurde nicht nur auf Uniformen von Eisenbahnangestellten aufgenäht, sondern auch von der Reifen- und Elektroindustrie adaptiert.

Ab 1959 finden wir das aus Stier und geflügeltem Rad bestehende Markenzeichen in abgewandelter Form auch auf dem ersten Mini, der damals zunächst Morris Mini-Minor hieß. Erst ab 1969 vermarktete der britische Mutterkonzern British Leyland den MINI als eigene Marke mit eigenem Signet. Als Motiv wählte man allerdings ein modernes, abstraktes Logo, das jeden Zusammenhang mit dem bisherigen Morris-Markenzeichen vermied.

Dieses Logo – auch Mini Shield genannt – prunkte bis 1990 in unterschiedlicher Ausführung auf der Motorhaube des klassischen Mini. Als das Unternehmen 1990 den Mini Cooper neu auflegte, erfolgte erstmals die Rückbesinnung auf das ursprüngliche Morris-Wappen. Es entstand ein neues Markenzeichen: ein verchromtes Rad mit stilisier-

DAS LOGO
/
THE LOGO

The MINI brand insignia has seen many changes over the years. Its roots go back to the late 1940s. William Morris, the founder and owner of the British vehicle manufacturing company Morris Motor Company, had been awarded a peerage for his achievements in the automotive industry and for his charitable work, and from 1938 he had been entitled to use the title Viscount Nuffield. His coat of arms was an Ox Gules over three Waves Azure – the symbol of the City of Oxford – surrounded by a winged wheel. To commemorate its founder, the Morris company based its logo on this coat of arms and used it to decorate the successful 1948 model Morris Minor among others.

The motif of the winged wheel follows in the tradition of the allegorical representations of trains, whose origins go back to the mid-19th century. We can see that this is an imagined combination of a wheel and wings, a remarkable mixture of symbols for speed representing travel by air and travel by land. The natural flight of birds and the wheel created by man combine into a symbolically charged figure signifying high speed. The emblem was sown onto the uniforms of train employees and adapted by the tyre and electrical industries.

From 1959, we also see a different version of the trademark consisting of a bull and a winged wheel on the Mini, at the time initially called the Morris Mini-Minor. From 1969, Austin and Morris's British parent company British Leyland began marketing the Mini as a separate brand with an individual signet for the first time. However, a modern, abstract logo was chosen as a motif, with no connection to the former Morris trademark.

Different versions of this logo – also called the Mini Shield – remained on the bonnet of the classic Mini until 1990. When the company re-introduced the Mini Cooper in 1990, it referred back to the original Morris coat of arms for the first time. This resulted in a new logo: a chrome wheel and stylised wings, with 'MINI COOPER' in red letters at its centre on a white background. The words were framed by a green laurel wreath. By contrast, all other Minis at this time carried a different emblem that represented the new ownership structure and was based on the Rover trademark.

ten Flügeln, in dessen Mitte auf weißem Grund die roten Lettern »MINI COOPER« stand. Ein grüner Lorbeerkranz rahmte den Schriftzug. Die übrigen Mini dieser Jahre wiesen hingegen ein Emblem auf, das den neuen Eigentumsverhältnissen Rechnung trug und sich am Markenzeichen von Rover orientierte.

Mit der Übernahme durch die BMW Group wurde die Markenidentität von MINI neu definiert und konsequent umgesetzt. Das aktuelle Markenzeichen baut auf dem in den Neunzigerjahren entwickelten Logo auf und zeigt ein in Chrom gehaltenes Rad mit stilisierten Flügeln. In seiner Mitte steht auf schwarzem Untergrund die Aufschrift MINI – eine moderne Interpretation des Vorgängers und eine Verbeugung vor einer langjährigen Tradition.

When the BMW Group took over in 1994, the brand identity of the MINI was re-defined and implemented consistently. The current trademark builds on the logo developed in the 1990s, consisting of a chrome wheel with stylised wings. At its centre, we see the word MINI on a black background – a modern interpretation of its predecessor and an acknowledgement of a long tradition.

FIRMENGESCHICHTE / COMPANY HISTORY

1913
William Richard Morris gründete die Morris Motor Company.

In den Zwanziger- bzw. Dreißigerjahren übernahm Morris die beiden Marken Riley und Wolseley. Gemeinsam bildeten sie die sogenannte Nuffield-Gruppe.
/
William Richard Morris founded the Morris Motor Company.

Morris took over the brands Riley and Wolseley in the 1920s and 1930s. Together, they became known as the Nuffield Group.

1952
Die Nuffield-Gruppe fusionierte mit Austin zur British

Motor Corporation (BMC). Die Leitung übernahm William Morris alias Lord Nuffield. Daimler und Jaguar schlossen sich an.
/
The Nuffield Group merged with Austin to become the British Motor Corporation (BMC). William Morris, later Lord Nuffield, took over the management. The Group was joined by Daimler and Jaguar.

1968
Die BMC vereinigte sich mit einem weiteren britischen Verbund von Automobilherstellern, der Leyland Motors. Diese Gruppe bestand aus dem Nutzfahrzeughersteller Leyland, der 1961 die Firma Standard-

Triumph und 1967 Rover erworben hatte. BMC und Leyland Motors fusionierten zur British Leyland Motor Corporation (BLMC).
/
The BMC merged with another British car manufacturing group, the Leyland Motors. This group consisted of the commercial vehicle manufacturer Leyland, which had acquired the Standard-Triumph car company in 1961 and Rover in 1967. BMC and Leyland Motors merged to become the British Leyland Motor Corporation (BLMC).

1975
Der Konzern wurde verstaatlicht.
/

The company was bailed out by the British Government.

1977
Der Konzern wurde zu British Leyland Ltd.
/
The company becomes known as British Leyland Ltd.

1986
Die British Leyland nannte sich fortan Rover Cars.
/
British Leyland changed its name to Rover Cars.

1988
Rover Cars wurde unter der Regentschaft der britischen Premierministerin Margaret Thatcher reprivatisiert und vom Luft- und Raumfahrt-

konzern British Aerospace aufgekauft.
/
Under British Prime Minister Margaret Thatcher Rover Group was re-privatised and bought by the aviation and space travel group British Aerospace.

1994
BMW übernahm den Konzern Rover Cars.
/
BMW took over the Rover Group.

2000
Nach hohen Verlusten verkaufte BMW den Konzern in zwei Teilen. Die Marke MINI blieb in seinem Besitz. Die PKW-Sparte, mittlerweile

MG Rover genannt, wurde von Phoenix übernommen. Landrover wurde von der Ford Motor Company aufgekauft.
/
After huge losses, BMW sold the group in two parts. It held on to the MINI brand, while the passenger vehicle division was sold to Phoenix, which re-named it MG Rover. The Ford Motor Company bought Land Rover.

Es gibt Marken, die mit ihrem Ursprungsland untrennbar verbunden sind. Ihre Heimat ist fester Bestandteil ihrer Identität und damit prägend für Charakter und Erscheinungsbild. Eines der Paradebeispiele hierfür ist der MINI, der in Großbritannien erfunden wurde, einem Schmelztiegel aus unterschiedlichen Kulturen, Denkweisen und Ideen, in dem besonders kreative und originelle Leistungen hervorgebracht wurden. Sir Alec Issigonis, der Erfinder des Mini, traf mit seiner Entwicklung den Puls der Zeit und überzeugte britische Trendsetter. Schnell etablierte sich der Mini in den Swinging Sixties und stieg von einem günstigen Massenkleinwagen zu einer Kultikone auf.

Im heutigen Zeitalter der Globalisierung vermitteln Marken mit lokalem Bezug ein eigenes Profil. Was sie auszeichnet, ist Originalität, Authentizität und Glaubwürdigkeit. MINI war im Lauf seiner wechselvollen Geschichte seit 1959 zwar in den Händen mehrerer Besitzer, dennoch stand der originäre Produktionsstandort am Rand der Industrie-

IDENTITÄT DER MARKE / BRAND IDENTITY

Some brands are inextricably linked with their country of origin. Their home is a fixed component of their identity, and a key influence on their character and appearance. The MINI is a prime example: invented in Britain, a melting pot of different cultures, ways of thinking and ideas, that has generated particularly creative and original achievements. Sir Alec Issigonis, the inventor of the Mini, had his finger on the pulse of the time and made an impact on British trendsetters. The Mini quickly became established in the Swinging Sixties, turning from an affordable mass-produced compact car into a cult icon.

Today, in the era of globalisation, brands with a local reference convey a very special profile. They are characterised by originality, authenticity and credibility. In the course of its varied history since 1959, MINI may have passed through the hands of several owners, but its original production site on the edge of the industrial metropolis of Birmingham was never in doubt. In 1994, when the BMW Group took

MINI INTERNATIONAL

Abu Dhabi / Abu Dhabi	Bulgarien / Bulgaria	Griechenland / Greece	Korea / Korea	Norwegen / Norway	Spanien / Spain
Ägypten / Egypt	Chile / Chile	Großbritannien / Great Britain	Kroatien / Croatia	Oman / Oman	Sri Lanka / Sri Lanka
Albanien / Albania	China / China	Guadeloupe / Guadeloupe	Kuwait / Kuwait	Österreich / Austria	St. Lucia / Saint Lucia
Algerien / Algeria	China Hongkong /	Guam / Guam	Lettland / Latvia	Panama / Panama	Südafrika / South Africa
Angola / Angola	China Hong Kong	Guatemala / Guatemala	Libanon / Lebanon	Paraguay / Paraguay	Tahiti / Tahiti
Argentinien / Argentina	China Taiwan / China Taiwan	Indien / India	Litauen / Lithuania	Peru / Peru	Thailand / Thailand
Armenien / Armenia	Costa Rica / Costa Rica	Indonesien / Indonesia	Malaysia / Malaysia	Philippinen / Philippines	Trinidad / Trinidad
Aruba / Aruba	Dänemark / Denmark	Irland / Ireland	Malta / Malta	Polen / Poland	Tschechien / Czech Republic
Aserbaidschan / Azerbaijan	Deutschland / Germany	Israel / Israel	Marokko / Morocco	Portugal / Portugal	Tunesien / Tunisia
Australien / Australia	Dominikanische Republik /	Italien / Italy	Martinique / Martinique	Réunion / Réunion	Türkei / Turkey
Bahamas / Bahamas	Dominican Republic	Jamaika / Jamaica	Mexiko / Mexico	Rumänien / Romania	Ukraine / Ukraine
Bahrain / Bahrain	Dubai / Dubai	Japan / Japan	Moldawien / Moldavia	Russland / Russia	Ungarn / Hungary
Barbados / Barbados	El Salvador / El Salvador	Jemen / Yemen	Montenegro / Montenegro	Saudi Arabien / Saudi Arabia	Uruguay / Uruguay
Belgien / Belgium	Elfenbeinküste / Ivory Coast	Jordanien / Jordan	Neukaledonien /	Schweden / Sweden	USA / USA
Bermudas / Bermuda	Estland / Estonia	Kanada / Canada	New Caledonia	Schweiz / Switzerland	Vietnam / Vietnam
Bolivien / Bolivia	Finnland / Finland	Kasachstan / Kazakhstan	Neuseeland / New Zealand	Senegal / Senegal	Weißrussland / Belarus
Bosnien-Herzegowina /	Frankreich / France	Katar / Qatar	Niederländische Antillen /	Serbien / Serbia	Zypern / Cyprus
Bosnia-Herzegovina	Französisch-Guyana /	Kaymaninseln /	Netherlands Antilles	Singapur / Singapore	
Brasilien / Brazil	French Guiana	Cayman Islands	Niederlande / Netherlands	Slowakei / Slovakia	
Brunei / Brunei	Georgien / Georgia	Kolumbien / Colombia	Nigeria / Nigeria	Slowenien / Slovenia	

metropole Birmingham nie in Frage. Auch 1994, als die BMW Group die Marke übernahm, war MINI ganz klar »made in Britain«. Die Wahrnehmung der Marke als kreativ, humorvoll, ironisch und exzentrisch ist eng mit der britischen Kultur und Wesensart verbunden, welche die Historie, den Namen, wie auch Philosophie und Charakter des MINI prägen.

Die Initiative und die erste Produktionsstätte sind im Vereinigten Königreich verortet, wo der MINI jahrzehntelang auch seinen größten Absatzmarkt hatte. Fans in aller Welt verbinden mit ihrem geliebten MINI ein originär britisches Produkt. Die festen Wurzeln in Longbridge beziehungsweise in Oxford behindern die internationale Ausrichtung des MINI jedoch keineswegs. Ein Grund dafür mag der British Commonwealth sein, der britische Erzeugnisse schon immer weltumspannend zirkulieren ließ. Bereits im Jahr 1959 startete parallel zur Produktion in England eine Lizenzfertigung in den Niederlanden. Es gehört zu den sympathischen Winkelzügen der Geschichte, dass sich dieser Schritt in der jüngsten Gegenwart wiederholt: Im Juli 2014 wurde im limburgischen Born ein Vertragswerk eröffnet, das den MINI in vielerlei Modellvarianten fertigt.

In den Sechziger- bis Achtzigerjahren exportierte die British Motor Corporation den MINI von England aus in 25 Länder. Frankreich, Japan und Westdeutschland galten damals als wichtigste Absatzmärkte. Um die weltweit steigende Nachfrage bedienen zu können, wurde der Wagen in mehreren Ländern direkt vor Ort zusammengebaut – seine relativ einfache Konstruktion machte dieses Verfahren möglich. Zu den international größten Mini-Herstellern zählten Belgien, Spanien, Südafrika, Chile, Venezuela, Australien und Frankreich. In Italien produzierte das Unternehmen Innocenti den Mini in Lizenz, in Portugal ging mit dem Moke ein eigenes Mini-Modell vom Band.

Heute findet der MINI – der klassische wie der neue – zahlreiche Freunde. Die Marke erfreut sich eines sehr hohen Sympathiewerts, unabhängig von Schichten, Altersgruppen und Nationalitäten. Mittlerweile ist sie in 110 Ländern dieser Erde vertreten. Zu den wichtigsten Absatzmärkten zählen die USA, Großbritannien, Deutschland, China, Italien, Frankreich, Japan, Spanien und Südkorea (Stand August 2014).

over the brand, Mini remained unquestionably 'made in Britain'. The brand is perceived as creative, humorous, ironic and eccentric and as such closely linked to British culture and way of life, all of which characterise the history and name, as well as the philosophy and character, of the MINI.

The initial project and its first production site are located in the United Kingdom, where MINI also had its biggest sales market for decades. Fans all over the world consider their beloved MINI to be an original British product. Nevertheless, its strong roots in Longbridge and Oxford are no obstacle to the MINI's international focus. One reason for this might be the British Commonwealth, which always circulated British products around the world. As early as 1959 licence production was launched in parallel in the Netherlands. And a happy turn of events in this story saw this repeated in recent history: in July 2014, a factory for contract work opened in Born, Limburg, where many different versions of the MINI are manufactured.

Between the 1960s and 1980s, the British Motor Corporation exported the Mini from England to 25 countries. France, Japan and West Germany were the most important sales markets at the time. In order to be able to meet rising global demand, the cars were assembled on site in several countries, a process facilitated by their relatively simple construction. Belgium, Spain, South Africa, Chile, Venezuela, Australia and France were among the biggest international manufacturers of the Mini. In Italy, the Innocenti company undertook licence production of the Mini; and in Portugal, a separate model of the Mini, the Moke, was released.

Today, both the classic and the new MINI maintain their popularity. The brand continues to attract attention, irrespective of social standing, age groups or nationalities. Today, it is sold in 110 countries. Currently its most important sales markets include primarily the USA, Great Britain, Germany, China, Italy, France, Japan, Spain and South Korea (as at August 2014).

Das Phänomen MINI ist im Stammbaum der britischen Automobil-industrie fest verankert, aber auch eng mit dem Namen seines Kon-strukteurs Sir Alec Issigonis verbunden.

1906 Am 18. November 1906 kam Alec Issigonis, eigentlich Alexander Arnold Constantine Issigonis, in Smyrna (dem heutigen Izmir) an der Ägäisküste zur Welt. Sein Vater Konstantin Issigonis besaß ein Unter-nehmen für Schiffsmotoren und konnte seinen Sohn schon früh für Eisenbahnen und Dampfmaschinen begeistern.

1919 Als Alec zwölf Jahre alt war, brach der griechisch-türkische Krieg aus. Türkische Soldaten eroberten Smyrna. Der politische Wandel sorgte für Turbulenzen, Smyrna stand damals unter griechischer Verwaltung. In der hauptsächlich von Türken bewohnten Hafenstadt herrschte eine explosive Stimmung. Der Zorn der Einheimischen richtete sich vor allem gegen die hier lebenden Griechen, die in der Minderheit waren.

1922 Die Türken eroberten das besetzte Westanatolien einschließlich Smyrna. Alecs Vater war Grieche und besaß die britische Staats-bürgerschaft, die Mutter Hulda Prokopp stammte aus einer württem-bergischen Brauereifamilie. Da es britischen Staatsbürgern nicht mehr möglich war, in Anatolien sicher zu leben, entschied sich die Familie für die Ausreise nach Malta mit dem Ziel der Weiterreise nach Groß-britannien.

Für Alecs Eltern war die Flucht von extremen Verlusten begleitet: Nicht nur, dass sie ihr gesamtes Vermögen in Smyrna zurücklassen mussten. Während der Überfahrt nach Malta erkrankte der Vater und blieb zu-rück, während Mutter und Sohn Anfang 1923 die strapaziöse Passage nach Dover auf sich nahmen. Kaum fort, erfuhren sie von dem Tod des Vaters.

1924 Alec nahm sein erstes eigenes Auto in Besitz: einen Singer mit Weymann-Karosserie.

1925 Ein Jahr später unternahm er gemeinsam mit seiner Mutter eine Autoreise durch halb Europa – eine »ununterbrochene Pannenserie«, wie er sich später erinnerte. Offenbar beeindruckt von den Möglich-

SIR ALEC ISSIGONIS

Rechts / Right
Sir Alec Issigonis an seinem Schreibtisch / Sir Alec Issigonis at his desk

The MINI phenomenon has firm roots in the family tree of the British automotive industry, but is also closely associated with the name of its designer, Sir Alec Issigonis.

1906 On 18 November 1906, Alec Issigonis (full name Alexander Arnold Constantine Issigonis), was born in Smyrna (today's Izmir) on the Aegean Coast. His father Konstantin Issigonis owned a ship engine business, and from early on shared his enthusiasm for trains and steam engines with his son.

1919 When Alec was 12 years old, the Greco-Turkish War broke out and Turkish soldiers conquered Smyrna. Political changes led to a tur-bulent period. At the time, Smyrna had a Greek administration. The mood in the harbour city, which was inhabited mainly by Turks, was explosive. The locals' anger was focused primarily on the Greek mi-nority living there.

1922 Turkey conquered occupied Western Anatolia, including Smyrna. Alec's father was Greek with British citizenship, while his mother Hulda Prokopp came from a Württemberg brewing family. Since it was no longer safe for British citizens to live in Anatolia, the family decided to set off for Malta with the aim of travelling on to Britain.

For Alec's parents, their escape involved devastating losses. Not only were they forced to leave all of their possessions in Smyrna, but on the boat to Malta his father became ill and had to remain behind, while mother and son continued on their arduous crossing to Dover alone at the start of 1923. They had barely left when they learnt of Konstantin's death.

1924 Alec bought his first car – a Singer with a Weymann body.

1925 A year later, he and his mother drove across half of Europe – an 'uninterrupted series of breakdowns', as he later recalled. Clearly fas-cinated by the sheer possibilities offered by technology, Alec enrolled at London's famous Battersea Polytechnic aged 19 to study mechanical engineering for three years. Though he achieved good results in his drawing lessons, he failed maths three times in a row. His talent in crafts and his enthusiasm for mechanical drawing were offset by his

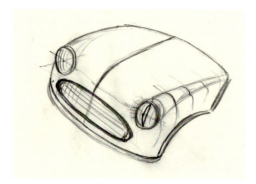

Entwurfsskizze zur Front / Design sketch for a bonnet
von / by Sir Alec Issigonis

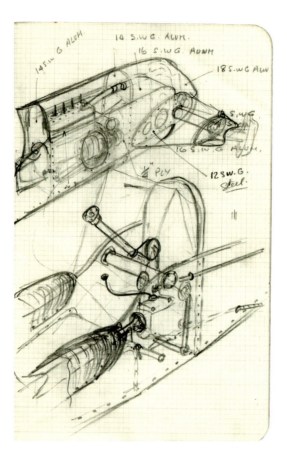

Entwurfsskizze für eine Leichtbaukonstruktion / Design sketch
for a lightweight construction von / by Sir Alec Issigonis

keiten der Technik, schrieb sich Alec mit 19 Jahren am renommierten Polytechnikum im Londoner Stadtteil Battersea ein, um eine dreijährige Ausbildung in Maschinenbau zu beginnen. Zwar lieferte er gute Arbeit im Fach Zeichnen ab, fiel aber in Mathematik dreimal durch. Seine Begabung fürs Handwerkliche und seine Begeisterung fürs Konstruktionszeichnen glichen seine ausgeprägte Abneigung gegen theoretische Mathematik aus. Er schaffte den Abschluss, ein weiterführender Studienplatz blieb ihm jedoch verwehrt. Also konzentrierte sich Alec Issigonis auf einen Posten als technischer Zeichner und fand bei Edward Gillet, einem Londoner Konstruktionsbüro für Automobiltechnik, eine Anstellung.

1934 Issigonis wechselte ins Konstruktionsteam des Automobilherstellers Humber Ltd. nach Coventry und beteiligte sich dort an der Einführung von Einzelradaufhängungen.

1936 Aufgrund seiner Kenntnisse in der Fahrzeugentwicklung warb ihn die Firma Morris Motors ab. In den Kriegsjahren beschäftigte sich Alec mit Militärfahrzeugen.

1941 Morris startete das Projekt Mosquito, einen kompakten Viersitzer, der nach dem Krieg eingesetzt werden sollte. Trotz widriger Umstände stellten Issigonis und sein Team in nur drei Jahren den ersten fahrfertigen Prototypen auf die Räder.

1948 Besser bekannt unter dem Namen Morris Minor wurde dieses Fahrzeug zu einem der erfolgreichsten Kleinmodelle der Nachkriegszeit.

1952 Morris und die Austin Motor Company fusionierten zur British Motor Corporation (BMC). Wegen mangelnder Zukunftsperspektiven wechselte Issigonis zum Autobauer Alvis, bei dem er eine Luxuslimousine entwickeln sollte. Doch scheiterte das Projekt an fehlenden finanziellen Mitteln.

1955 Der Vorstandsvorsitzende von BMC, Leonard Lord, holte Issigonis als stellvertretenden Technischen Leiter ins Austin-Werk nach Longbridge zurück, wo er neue Baureihen für die Kleinwagen-, die Mittelund Oberklasse entwickeln sollte. Im Zentrum der Überlegungen: der neue Austin 1500.

extreme dislike of theoretical mathematics. Nevertheless he finished his degree, but was unable to secure a place for further studies. Issigonis decided to become a technical draughtsman and found employment with Edward Gillet, a design office for automotive technology in London.

1934 Issigonis moved to the design team of the car manufacturer Humber Ltd. in Coventry, where he was involved in introducing independent wheel suspensions.

1936 Given his knowledge of vehicle development, he was headhunted by Morris Motors. During the war, Alec focused on military vehicles.

1941 Morris launched the Mosquito project, a compact four-seater to be released after the war. In spite of the difficult circumstances, Issigonis and his team produced the first roadworthy prototype in just three years.

1948 Better known as the Morris Minor, this car became one of the most successful compact models of the post-war era.

1952 Morris and the Austin Motor Company merged to form the British Motor Corporation (BMC). Due to a lack of job prospects, Issigonis moved to car manufacturer Alvis, where he was tasked with develop-

Doch nahmen die Ereignisse eine unerwartete Wende: In Ägypten änderten sich die politischen Verhältnisse, was spürbare Konsequenzen für Großbritannien haben sollte. Bereits Jahre zuvor war es in Nordafrika zu schweren Unruhen gekommen: Nationale Bewegungen kämpften gegen die koloniale Herrschaft der Franzosen. Großbritannien hingegen zog sich schrittweise aus dem Nahen Osten zurück und war bereit, die Unabhängigkeit Ägyptens unter der Herrschaft von König Faruq anzuerkennen. Doch Truppeneinheiten der ägyptischen Armee stürzten den Monarchen. An die Spitze des neuen Regimes stellte sich Präsident Gamal Abdel Nasser. Angesichts der schlechten Wirtschaftslage seines Landes ließ er am 26. Juli 1956 die Suezkanal-Gesellschaft, die in ausländischer Hand lag, verstaatlichen. Mit den Jahren war die wirtschaftliche Bedeutung des Erdöls evident geworden, ebenso die Abhängigkeit der europäischen Industrienationen von der freien Fahrt durch den Suezkanal, weshalb englische und französische Truppen in das Geschehen eingriffen. Doch die Angst vor einer Eskalation mitten im Kalten Krieg bewog die Europäer schließlich zum Abzug.

In Großbritannien war man bestürzt über den Verlust des Einflusses auf Ägypten und schockiert durch die Tatsache, dass die Suezkrise fast die gesamte Treibstoffproduktion des Landes lahmlegte. Bis dahin waren fast zwei Drittel des importierten Rohöls über den Suezkanal transportiert worden. Die Regierung griff zu rigorosen Maßnahmen und beschloss die Rationierung der Spritabgabe. Dies führte zu einem Ansturm auf verbrauchsgünstige Kleinwagen, sodass der Vorstand von BMC dem Austin 1500, der damals entwickelt wurde, kaum noch Chancen einräumte. Ein neuer Kleinwagen musste her!

1956 Leonard Percy Lord, auch als Baron Lambury bekannt und seines Zeichens Chef von British Motor Corporation, forderte Issigonis auf, einen revolutionären Kleinwagen zu entwickeln, der auf einem Minimum an Raum ein Maximum an Platz bieten sollte.

1957 Nur sieben Monate später verkündete Issigonis die Konstruktion zweier Prototypen und lud Leonard Lord zu einer Testfahrt ein. Später berichtete er: »Wir fuhren um das Werk, und ich raste mit ihm in einem

ing a luxury saloon car. The project failed, however, due to a lack of funds.

1955 The Chairman of the Board of BMC, Leonard Lord, brought Issigonis back as the Deputy Technology Officer at the Austin Works in Longbridge, where he was to develop new series of compact cars as well as medium- and luxury-class cars. At the heart of these was the new Austin 1500.

Events took an unexpected turn, however. Political circumstances changed in Egypt, and this was to have tangible consequences for Britain. Years earlier, North Africa had experienced serious unrest, as nationalist movements fought against French colonial domination. By contrast, Britain was gradually withdrawing from the Middle East and was willing to recognize Egypt's independence under the reign of King Farouk. However, military units of the Egyptian Army overthrew the monarch. The new regime was led by President Gamal Abdel Nasser. Given the country's unfavourable economic situation, he nationalised the foreign-owned Suez Canal Company on 26 July 1956. Over the years, the economic importance of oil had become apparent, as had the dependence of European industrial nations on free passage through the Suez Canal, with the result that English and French troops intervened. But the fear of escalation in the middle of the Cold War finally convinced the Europeans to withdraw.

Britain was distraught at the loss of influence over Egypt and shocked by the fact that the Suez Crisis disabled almost the entire fuel production of the country. Until then, nearly two thirds of imported crude oil had been transported via the Suez Canal. The government took rigorous measures and decided to ration fuel. This led to a rush to buy economical compact cars with the result that the Board of Directors at BMC saw few opportunities for the Austin 1500, which was then in development. There was a dire need for a new baby car.

1956 Leonard Percy Lord, 1st Baron Lambury, the then president of the British Motor Corporation, asked Issigonis to develop a revolutionary baby car that offered a maximum interior space while taking up only minimal room on the road.

Sir Alec Issigonis am Mini-Produktionsstandort Longbridge /
at the Mini production plant in Longbridge

Höllentempo. Ich bin sicher, dass er Angst hatte, aber dann war er sehr beeindruckt von der Straßenlage. Wir hielten vor seinem Büro. Er stieg aus und sagte: ›Los, bauen Sie dieses Auto!‹«

Issigonis war zu diesem Zeitpunkt ein anerkannter Konstrukteur mit unkonventionellem Werdegang und außergewöhnlichen Arbeitsmethoden. Zeitgenossen beschreiben ihn als akribischen Tüftler und Techniker aus Leidenschaft. Er zeichnete seine Entwürfe seltener am Reißbrett, sondern mit Vorliebe auf seinen stets griffbereiten Skizzenblock.

1959 Am 26. August wurde das ersehnte neue Modell als Morris Mini-Minor und als Austin Seven präsentiert. Der große Erfolg des Mini machte auch seinen Konstrukteur weltweit bekannt.

1961 Alec Issigonis wurde zum Technischen Direktor von BMC berufen und Mitglied im Vorstand der Austin Motor Company.

1963 Issigonis übernahm den Vorsitz der gesamten British Motor Company.

1967 Die Royal Society, Großbritanniens angesehenste Forschungsvereinigung, nahm Issigonis als Mitglied auf.

1969 Queen Elisabeth II. schlug Issigonis zum Ritter.

1971 Sir Alec Issigonis trat in den Ruhestand, blieb dem Unternehmen aber bis 1987 als Berater treu.

1988 Issigonis starb am 2. Oktober, kurz vor seinem 82. Geburtstag.

1957 Only seven months later, Issigonis announced that he had built two prototypes and invited Leonard Lord to go on a test drive. Later Issigonis said: 'We drove around the works, and I was going at breakneck speed. I am sure that he was afraid, but then he was also very impressed by the road-holding. We stopped in front of his office. He got out and said: "Go on, build this car!"'

At the time, Issigonis was a respected designer with an unconventional career and extraordinary working methods. Contemporaries described him as a meticulous inventor and a passionate technician. He rarely made his sketches at a drawing board, but preferred using his ever-accessible sketch-pad.

1959 On 26 August, the eagerly awaited new model was presented as the Morris Mini-Minor and as the Austin Seven. The great success of the Mini also meant worldwide fame for its designer.

1961 Alec Issigonis was appointed Chief Technology Officer at BMC and became a member of the Board of Directors of the Austin Motor Company.

1963 Issigonis became the Engineering Director of the entire British Motor Company.

1967 He became a member of the Royal Society, Britain's most renowned research association.

1969 Issigonis was knighted by Queen Elizabeth II.

1971 Sir Alec Issigonis retired, but continued working for the company as an advisor until 1987.

1988 Issigonis died on 2 October 1988, shortly before his 82nd birthday.

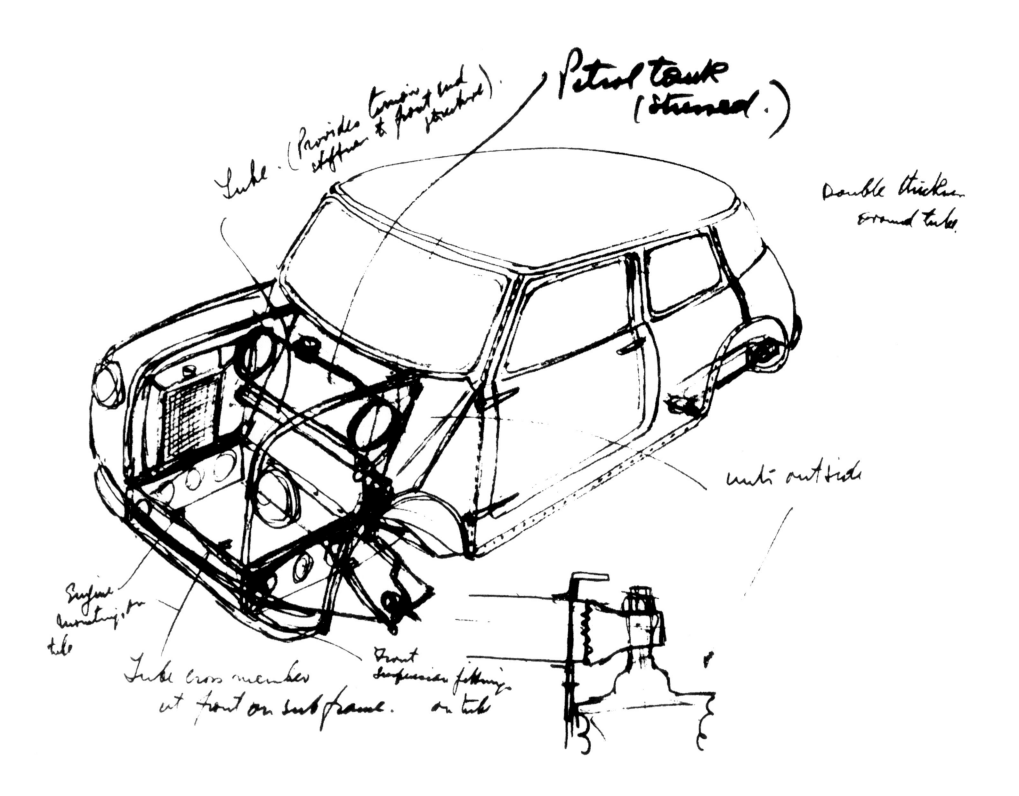

Handzeichnung von Sir Alec Issigonis für den Mini Prototyp / Sir Alec Issigonis's drawing for the prototype Mini, 1958

Die Vorgaben der Konzernleitung 1956 waren klar und unmissverständlich: viel Innenraum bei möglichst geringen Außenmaßen, Platz für vier Insassen, tadellose Fahreigenschaften, ein geringer Benzinverbrauch und ein günstiger Preis. Der Konstrukteur Issigonis hatte also einen sparsamen, viersitzigen Kleinwagen mit optimaler Raumausnutzung zu entwerfen. Vergleichbare Konzepte gab es nicht. Die technischen Herausforderungen verlangten einen grundlegend neuen Weg. Dabei hatte Issigonis freie Hand und wurde unterstützt durch ein Expertenteam, bestehend aus Jack Daniels, Chris Kingham, Charles Griffin, John Sheppard, Vic Everton, Ron Dovey, Dick Gallimore und George Cooper. Das erklärte Ziel war die Konstruktion eines »richtigen Kleinwagens«. In späteren Interviews bemerkte Issigonis, er hasse alles, was groß ist – große Häuser, große Firmen und – vor allem – große Autos. Der Kern seiner bahnbrechenden Neukonstruktion war das Konzept des Frontantriebs mit querstehendem Motor und unmittelbar darunter angeordnetem Getriebe. Beide Bauteile konnten mit einem gemeinsamen Ölkreislauf versorgt werden. In dem überschaubaren Motorraum fanden auch Kühler, Lenkung und Nebenaggregate Platz.

Das Prinzip des Frontantriebs in Verbindung mit dem »east-west«-Motor, also dem Einbau quer zur Vorderachse, bildete den Grundstein für die Kompaktheit des Mini und sollte zur Standardlösung moderner Klein- und Kompaktwagen werden.

Die extrem kurzen Überhänge der Karosserie, die Platzierung eines Rades an jeder Ecke und der daraus resultierende lange Radstand sowie der tiefe Schwerpunkt bildeten ideale Voraussetzungen für ein sehr agiles Fahrverhalten, das als Gokart-Feeling für MINI sprichwörtlich werden sollte. Optimale Raumökonomie und maximaler Fahrspaß wurden im klassischen Mini erstmals idealtypisch umgesetzt.

Um die Belastungen der leichten, selbsttragenden Stahlkarosserie zu verringern, lagerte das Konstruktionsteam den gesamten Antriebsstrang, die Lenkung und Aufhängung in einem Hilfsrahmen. Auch die hinteren Einzelräder waren an einem Hilfsrahmen befestigt, was dem Mini eine hervorragende Spurtreue bescherte. Doch auch die anderen Bestandteile des Fahrwerks boten eine Vielzahl technischer Meister-

DER ERSTE MINI / THE FIRST MINI

Links / Left
Präsentation des ersten
Austin Seven / Launch of the
first Austin Seven, 1959

In 1956 the management's specifications had been unmistakably clear: a large interior plus small exterior dimensions, space for four passengers, impeccable handling, low fuel consumption and an attractive price. So the task of designer Issigonis was to create an economical, compact four-seater, which made optimal use of space. There were no comparable concepts and the technical challenges required a totally new approach. Issigonis was given a completely free hand and was supported by a team of experts consisting of Jack Daniels, Chris Kingham, Charles Griffin, John Sheppard, Vic Everton, Ron Dovey, Dick Gallimore and George Cooper. The stated aim was to design a 'genuinely compact car'. In subsequent interviews, Issigonis commented that he hated everything large – large houses, large companies and, above all, large cars. At the heart of his ground-breaking new design was the concept of front-wheel drive with a transverse engine and the transmission located immediately below it in an expanded sump. Both components would use the same oil supply. The clearly understandable engine compartment also housed the radiator and steering system as well as ancillary components.

The principle of front-wheel drive in conjunction with the 'east-west' engine, i.e. its installation at right angles to the front axle, formed the basic building block for the Mini's compact size and was to become the standard solution for modern small and compact cars.

The extremely short front and rear overhang, with a wheel placed at each corner and the resulting long wheelbase and low centre of gravity, were the ideal conditions for very nimble handling that would create the go-kart feeling that epitomized the MINI. Optimal spatial economy and maximum driving pleasure found their first ideal expression in the classic Mini.

In order to reduce the pressure on the light, steel monocoque, the team of designers positioned the entire power-train, steering and suspension on separate front and rear sub-frames. This chassis configuration was another factor in the Mini's excellent directional stability. Other components of the chassis also included a number of technical achievements: in the Mini, coil, torsion or compound springs were

leistungen: Anstelle von Schrauben-, Torsions- oder Blattfedern erhielt der Mini eine Gummifederung. Dazu diente ein Gebilde aus zwei Kegeln mit einer dazwischenliegenden Gummischicht. Der obere Kegel war fest mit dem Hilfsrahmen verschraubt, der untere mit dem Radträger. Da sich Gummi mit zunehmendem Druck verhärtet, erhielt der Wagen somit eine progressive Federung. Dieses System hatte so gute Eigenschaften, dass klein dimensionierte Teleskopstoßdämpfer ausreichten. Um ein möglichst feines Ansprechverhalten zu erzielen, waren sie außen an den oberen Querlenkern vorn und den hinteren Längslenkern befestigt.

Auch bei der Kraftübertragung wurden neue Wege beschritten. Weil die bis dahin üblichen Kardangelenke bei größeren Lenkeinschlägen zum Verziehen neigten, entschied sich Issigonis dafür, erstmals in einem Automobil homokinetische Gelenke einzusetzen. Sie bestanden aus einem Kugellager, das von drei Käfigen umschlossen war, von denen zwei mit dem An- und dem Abtrieb verbunden waren. Diese Konstruktion erlaubte ausreichende Lenkwinkel ohne Verwindungen und reduzierte die Antriebseinflüsse auf die Lenkung ganz erheblich.

Mit einer Länge von 3,05 Metern und einer Breite von 1,41 Metern waren die Außenmaße des Mini relativ frühzeitig in der Entwicklung definiert worden. Sein Leergewicht betrug nicht einmal 600 Kilogramm. Möglich wurde dies nur durch die eiserne Disziplin von Issigonis und seinen Mitarbeitern: Rund 80 Prozent des Volumens, das der Kleinwagen beansprucht, stehen den Passagieren und ihrem Gepäck zur Verfügung. Es gibt vier vollwertige Sitzplätze und einen Kofferraum mit 195 Litern Fassungsvermögen. Entsprechend minimalistisch fiel die Gestaltung der Fahrgastzelle aus: Zum Öffnen der Tür diente ein schlichter Seilzug. Vor Fahrer und Beifahrer spannte sich statt eines Armaturenbretts eine Ablage. In ihrer Mitte saß als Zentralinstrument der Tacho mit Meilenzähler und Benzinuhr, darunter zwei Kippschalter für Scheibenwischer und Licht. Eine Heizung gab es nur gegen Aufpreis.

Bei seiner Premiere trug der Kleinwagen noch nicht den Namen »Mini«. Offiziell kam er fast modellgleich mit den Bezeichnungen Mor-

replaced by rubber suspension. This was a structure consisting of two cones separated by a layer of rubber. The upper cone was screwed on to the auxiliary chassis, while the lower one was attached to the wheel bearing. Since rubber hardens under increased pressure, this resulted in the car's progressive suspension system, which worked so well that only small, telescopic shock absorbers were required. In order to achieve the greatest degree of responsiveness, these shock absorbers were attached to the upper control arms at the front and to the rear trailing-arm suspension at the back.

The Mini also broke new ground in terms of its transmission. Since the universal joints most often employed until that time tended to distort when operated at more extreme angles, Issigonis decided to install homokinetic, constant-velocity joints in a car for the first time. These consisted of ball bearings surrounded by three cages, two of which were linked to the input and output shafts. This design meant that a range of steering angles could be achieved without stress, thus significantly reducing the impact of the drive shaft on steering.

Measuring 3.05 m long by 1.41 m wide, the external dimensions of the Mini had been defined relatively early on in its development. Unladen its weight was less than 600 kg. This had come about thanks only to the steely determination of Issigonis and his team: around 80% of this compact car's volume is available for the passengers and their luggage. There are four full-sized seats and a 195-litre capacity boot. The design of the passenger compartment was similarly minimalist. For example, the door opened from inside by pulling a simple wire, and instead of a dashboard, the driver and front passenger sat facing a storage rack. At its centre were the speedometer with mile counter and fuel gauge, and below them two rocker switches for the windscreen wipers and lights. Heating was only available at an extra cost.

When it was first presented, this compact vehicle was not yet called a 'Mini'. Officially, two practically identical models were released on the market under the names Morris Mini-Minor and Austin Seven, thus enabling the British Motor Corporation (BMC) to exploit the broad range of its brand.

Der erste »MINI« – ein Morris Mini-Minor / The first 'MINI' – a Morris Mini-Minor, 1959

ris Mini-Minor und Austin Seven auf den Markt. Damit trug die British Motor Corporation (BMC) ihrem breiten Markenspektrum Rechnung.

Am 3. April 1959 lief das erste Modell des Austin Seven neuen Zuschnitts im Austin-Werk Longbridge vom Band. Mit der Wahl des Namens baute die British Motor Corporation bewusst auf Tradition, denn in den Jahren 1922 bis 1939 hatte England einen legendären Austin Seven fast 300 000-mal gebaut und mit ihm die Massenmotorisierung in Großbritannien eingeläutet. An diesen Erfolg sollte das kleine Fahrzeug 1959 anknüpfen.

Auch die Marke Morris bot ihren Kunden den baugleichen Viersitzer als Morris Mini-Minor an. Dieser verließ das Morris-Werk in Oxford nur wenige Wochen später, am 8. Mai 1959. Gemeinsam wurden beide Modelle am 26. August 1959 der Öffentlichkeit präsentiert. Trotz ihrer unterschiedlichen Herkunft waren beide Modelle fast identisch. Die Unterschiede beschränkten sich auf äußere Besonderheiten, auf den Kühlergrill, die Radkappen sowie die Karosseriefarben. Der Austin Seven war in den Farben Tartan Red, Speedwell Blue und Farina Grey erhältlich, für den Morris Mini-Minor standen Lackierungen in Cherry Red, Clipper Blue und Old English White zur Auswahl.

Die internationale Fachpresse berichtete wohlwollend über den neuen Kleinwagen. Die deutsche *Motor Revue* schrieb 1960: »Dieser Wunderwagen (Gummifederung, quer gestellter Vierzylindermotor, Motor und Getriebe in einem Ölsumpf, billige 10-Zoll-Reifen, ungewöhnlich großer Innenraum, kleine Verkehrsfläche) hätte es anders verdient, denn bei uns wird oft für mehr Geld Unvollkommeneres gekauft – aber unserem Käuferpublikum fehlt der klare Blick.«

Tatsächlich aber kam der Absatz des neuen Kleinwagens – made in Britain – nicht auf Touren. Für junge Käufer war er trotz seines günstigen Preises von 496,95 Pfund zu teuer, für Besserverdienende hingegen fiel die Ausstattung zu spartanisch aus.

On 3 April 1959, the first model of the new-look Austin Seven rolled off the assembly line at the Austin factory in Longbridge. In selecting its name, the British Motor Corporation was consciously building on tradition, as between 1922 and 1939 England had produced nearly 300,000 legendary Austin Sevens, thus heralding mass car-ownership in Great Britain. In 1959, the small car was to continue this success.

The Morris brand also offered its customers the similarly built four-seater Morris Mini-Minor. This left the Morris works in Oxford only a few weeks later on 8 May 1959. Both models were presented to the public together on 26 August 1959. In spite of their different origins, the two models were almost identical. Their differences were limited to external features, such as the radiator grill, hub-caps and bodywork colours. The Austin Seven was available in the colours of Tartan Red, Speedwell Blue and Farina Grey, while the Morris Mini-Minor could be ordered in Cherry Red, Clipper Blue and Old English White.

The international trade press reported favourably on the new small car. The German *Motor Revue* wrote in 1960: 'This miracle car (rubber suspension, transverse four-cylinder engine, engine and gear-box-in-sump, affordable 10-inch tyres, unusually large passenger compartment, small on-road size) should have deserved more recognition, given that here [in Germany] more is often spent on less – but our buying public lacks a clear understanding'.

In the end, however, sales of this new small car – made in Britain – really did struggle. In spite of its affordable price of £496.95, it was too expensive for young buyers; and higher-income earners considered its features too spartan.

Die Zielsetzung – ein Minimum an äußerer Dimension bei einem Maximum an Platzangebot – waren maßgeblich für das Erscheinungsbild des neuen, revolutionären Kleinwagens. In der Tat war der Mini von Beginn an ein Raumwunder. So viel Platz hatte kein Automobil vor ihm zu bieten: 80 Prozent des gesamten Fahrzeugvolumens waren den Insassen und ihrem Gepäck vorbehalten. Ein in den frühen Sechzigerjahren produziertes Schnittmodell demonstriert das geniale Konzept und die darauf abgestimmte Gesamtkonfiguration des Wagens. Aus einer der ersten Konstruktionsphasen stammt eine kolorierte Schnittzeichnung, die das geradezu »intelligente« Platzangebot unterstreicht. Die drei hier eingezeichneten Gepäckstücke im Kofferraum dürften allerdings nur »Miniformat« gehabt haben.

Auch in der zeitgenössischen Werbung – ob als kolorierte Zeichnung oder Schwarz-Weiß-Fotografie – wird der Komfort, den mehrere Fahrgäste hinsichtlich Zuladung genießen, geradezu inszeniert. Augenfällig ist die erstaunliche Fülle an Gepäck, die im Mini Platz findet.

Möglich wurde das sensationelle Raumangebot durch die Platzierung der vier Räder an den äußersten vier Ecken und die Verwendung kleiner, eigens entwickelter 10-Zoll-Räder. Die Reifenindustrie hatte damals generell größere Reifen mit 16 oder 17 Zoll Durchmesser im Programm. Issigonis setzte sich mit Experten der Firma Dunlop zusammen und konnte sie überzeugen, kleinere Reifen ins Angebot aufzunehmen.

Der Kofferraum fasste 195 Liter. Wem das nicht ausreichte, der ließ einfach die Klappe offen und legte Gepäckstücke aller Art darauf ab. Da die Klappe unten angeschlagen war, konnten sogar sperrigere Güter darauf deponiert und einigermaßen sicher befestigt werden. In dieser Handhabung sah man keine Verlegenheitslösung. Ganz im Gegenteil: In glänzenden Verkaufsprospekten wurde die erweiterte Ladekapazität farbenfroh präsentiert.

Das beeindruckende Platzangebot führte im Lauf der Jahre zu Experimenten mit sportlichem, ja geradezu artistischem Anspruch. Öffentlichkeitswirksam wurden vor allem junge Damen – Studentinnen, Mannequins, Sportlerinnen – dazu animiert, sich kunstvoll im

EIN RAUMWUNDER / A MIRACLE OF SPACE

Links / Left
Morris Mini Traveller, 1961

The targets – a maximum of interior space with minimum external dimensions – were key in shaping the look of the new, revolutionary small car. In fact, from the start the Mini was a miracle of space. No car had previously offered so much room in a given space: 80% of the entire vehicle volume was dedicated to the occupants and their luggage. A cutaway model produced in the early 1960s demonstrates this brilliant concept and the way the car's overall configuration is tailored to it. A coloured cutaway drawing from one of the first design phases emphasizes the almost 'intelligent' availability of space. However, the three items of luggage illustrated in the boot could only have been miniature in size.

Contemporary advertisements – whether colour illustrations or black-and-white photographs – also showcase the comfort that several passengers can enjoy once inside the car. The astonishing amount of luggage that can fit into a Mini is striking.

Locating the four wheels out the outside of the four corners and using small, specially developed 10-inch wheels enabled the vehicles' incredible spaciousness. At the time, the tyre industry generally sup-

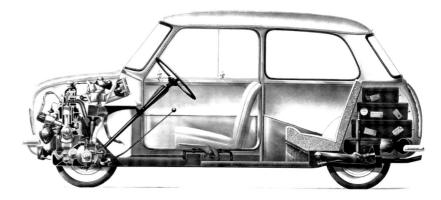

Längsschnitt des Mini / Longitudinal section of the Mini, 1959

Blick in den Kofferraum des MINI Clubman /
View of the MINI Clubman's boot space

Innenraum des MINI unterzubringen. Angesteckt vom Ehrgeiz der Olympioniken während der Olympischen Spiele 2004 in Athen pressten sich 21 erwachsene Menschen in einen klassischen Mini Cooper und brachen damit den bisherigen Rekord von 20 Personen.

Die Messlatte wurde 2014 deutlich höher gelegt, als sich die Gruppe Dani and the MINI-Skirts – mit insgesamt 27 Personen – am 18. Mai 2014 anlässlich des MINI-Rennens London–Brighton in einen klassischen MINI zwängte. Den Rekord beim neuen MINI halten 28 höchst gelenkige Turnerinnen, die 2014 im Londoner Potters Fields Park unter Beweis stellten, dass sie alle in einem Mini Cooper Platz fanden. Die beiden Rekorde von 2014 sind offiziell anerkannt und im *Guinness-Buch der Rekorde* vermerkt.

plied larger-sized 16- or 17-inch tyres. Issigonis sat down with experts at Dunlop and was able to convince them to include smaller tyres in their range.

The boot space had a capacity of 195 litres – and anyone who wanted even more could simply leave the tailgate open and put their luggage on it. Since the tailgate was hinged at the bottom, even outsized items could be positioned and secured there relatively safely. And this solution was not seen as a stopgap: on the contrary, glossy sales brochures showed colourful images of this increased load capacity.

Over the years, this impressive amount of space led to experimentation of a sporty, even artistic kind. Marketing campaigns saw young women in particular – whether students, models or athletes – being encouraged to pose artistically in the Mini's interior. Inspired by Olympian ambitions 21 adults squeezed into a classic Mini Cooper during the 2004 Olympics in Athens, thus breaking the previous record of 20 people.

The benchmark was set even higher in 2014 when on 18 May, on the occasion of the MINI race between London and Brighton, a group of 27 people calling themselves Dani and the MINI Skirts squeezed into a classic Mini. The record for the new MINI is now held by 28 extremely flexible female gymnasts, who in 2014 at Potters Fields Park in London proved that they could all fit into a MINI Cooper. The two 2014 records are officially recognized and included in the *Guinness Book of Records*.

»Dani and the MINI-Skirts« sichern sich 2014 den neuen Weltrekord /
'Dani and the MINI-Skirts' establish a new world record in 2014

Das Mini-Raumwunder / The Mini – a miracle of space

Restaurationsprozess des Mini Austin Seven / Restoring a Mini Austin Seven, 2013

Die Nachricht glich einer kleinen Sensation: 2012 war ein klassischer Mini – genauer gesagt ein Austin Seven – in einer Scheune im niederländischen Groningen gefunden worden. Er befand sich in einem desolaten Zustand, weshalb ein Jahr später Mitarbeiter der niederländischen Firma VDL Nedcar begannen, den Wagen fachgerecht zu zerlegen und zu restaurieren. Das Team setzte sich zusammen aus einem Karosseriebauer, einem Schweißer, einem Lackierer, einem Motorenspezialisten sowie einem Projektmanager. Der 34 PS starke Vierzylindermotor und das Getriebe wurden komplett auseinandergenommen und erneuert, einige Karosserieteile wie auch die Türverkleidungen von Hand nachgefertigt. Wo möglich, wurden originalgetreue Teile eingesetzt, ansonsten half die holländische MINI-Clubszene aus. In nur sechs Monaten konnte die aufwendige Restaurierung abgeschlossen werden. Heute erstrahlt der Mini mit der Nummer 983, auch Nedcar genannt, wieder in seiner ursprünglichen Lackfarbe Farina Grey. Die Nachforschungen ergaben, dass er 1959 als einer der ersten dreißig Mini in den Niederlanden montiert worden war. Fahrgestell- und Produktionsnummer lassen an der Herkunft keinen Zweifel. Damit ist der Wagen heute möglicherweise der älteste, noch fahrbereite klassische Mini aus niederländischer Fertigung. Zu den Ausführungen, die nur in den Niederlanden angeboten waren, gehört die Sitzpolsterung mit echtem Rosshaar.

Kurz nach Produktionsbeginn des Mini in England im August 1959 war die Entscheidung gefallen, Bauteile an die J. J. Molenaar's Car Companies im niederländischen Amersfoort zu exportieren. Dort entstanden zwischen 1959 und 1966 insgesamt mehr als 4000 Fahrzeuge – eine Leistung, welche für die erfolgreiche britisch-niederländische Kooperation im Fahrzeugbau spricht.

Diese Verbindung wurde kürzlich neu belebt: Seit Juli 2014 läuft bei dem Vertragsproduzenten VDL Nedcar im limburgischen Born der neue MINI vom Band. Die meisten Teile werden aus dem britischen Produktionsdreieck Oxford – Swindon – Hams Hall angeliefert.

DE EERSTE MINI

It was sensational news. In 2012, a classic Mini – an Austin Seven, to be precise – was found in a shed in Groningen in the Netherlands. It was in a sorry state, and so a year later employees of the Dutch company VDL Nedcar began to strip the car down professionally in order to restore it. The team consisted of a coachbuilder, a welder, a lacquerer, an engine specialist and a project manager. The 34 bhp, four-cylinder engine and gear-box were completely dismantled and updated, with some parts of the body and door panels re-built by hand. Where possible, original parts were used, failing which members of the Dutch MINI Club helped out. In just six months, the extensive restoration was complete. Today, Mini no. 983, also called Nedcar, is again resplendent in its original Farina Grey lacquer. Research revealed that in 1959 it was one of the first 30 Minis to have been assembled in the Netherlands. Chassis and production numbers leave no doubts as to its origins. Today this car is perhaps the oldest still drivable classic Mini constructed in the Netherlands. Among its features, which were available only in the Netherlands, is the genuine horsehair seat upholstery.

Shortly after production of the Mini was launched in Britain in August 1959, it was decided to export components to J. J. Molenaar's Car Companies in Amersfoort in the Netherlands. Between 1959 and 1966 more than 4,000 cars were made there – an achievement that demonstrates the success of Anglo-Dutch co-operation in vehicle construction.

This connection recently enjoyed a revival. Since July 2014 contract manufacturer VDL Nedcar in Born, Limburg, have been assembling the new MINI. Most of the parts are supplied by the British production triangle of Oxford – Swindon – Hams Hall.

Um zu ermessen, welche Bedeutung der MINI für die Automobilgeschichte ab 1959 hatte, empfiehlt sich zunächst ein Blick auf Schwarz-Weiß-Fotografien dieser Zeit. Auf Londons Straßen verkehrten dunkle, voluminöse Transportmittel – die den bis dato üblichen althergebrachten Automobilkonzepten entsprachen. Als der Mini als erster Kleinwagen auf den Markt kam, sputeten sich viele Autobauer, dessen Konzeption zu übernehmen. Mit seinem quer eingebauten Motor, dem Frontantrieb und den verblüffenden Außenmaßen war der Mini von nun an im Segment moderner Kleinwagen das Maß aller Dinge.

In Großbritannien gab es ein begrenztes Angebot an Kleinwagen. Der Verbraucher konnte wählen zwischen einer BMW Isetta für 390 Pfund, dem skurrilen Heinkel Kabinenroller für 399 Pfund oder dem Goggomobil T300 für 495 Pfund. Alle drei Modelle wurden aus Westdeutschland importiert, besaßen Hinterradantrieb und einen äußerst lauten Motorradmotor. Anschaffungspreis und Unterhalt waren erschwinglicher als bei »regulären« Autos. Nicht wenige Briten entschieden sich für einen Austin A35, einen Morris Minor 1000 oder einen Standard Eight. Diese erreichten immerhin 100 km/h und verbrauchten um die 9 Liter auf 100 Kilometern. Für diese Vorzüge nahm man eine harte Federung, ein äußerst beengtes Platzangebot für vier Personen sowie eine schlechte Straßenlage in Kauf.

Will man also den Stellenwert des MINI bewerten, waren Kriterien wie diese nicht unerheblich. Eines der beiden ersten Modelle des Mini, der Austin Seven, kostete stolze 496 Pfund. Für einen Citroën 2CV musste man hingegen tiefer in die Tasche greifen und 564 Pfund auf den Tisch legen, für einen Fiat 600 613 Pfund und für einen VW Käfer 617 Pfund. Mit seinem günstigen Preis entsprach der Mini der Vision eines Sir Alec Issigonis, der sich ursprünglich einen Mini vor jedem Arbeiterhaus erträumt hatte. Ob Prognose oder Wunschbild – der Wagen konnte die Verkaufserwartungen in den Sechzigerjahren nicht erfüllen.

Nicht anders gestaltete sich die Situation in Deutschland: »...aufräumen konnte man mit der Vergangenheit, die dem Fortschritt schon immer im Wege gestanden hatte, und zum Vollzug der Konsumgesellschaft schreiten, die in den Zwanzigerjahren Gestalt angenommen

DIE WELT DER KLEINWAGEN / THE WORLD OF COMPACT CARS

To get an idea of the importance of the MINI in the history of cars post-1959, it is worth looking at black-and-white photographs of the period. In those days the streets of London were full of dark, bulky vehicles, echoing the usual conventional, old-style layouts for cars. When the Mini hit the market as a pioneering four-seater compact car, many car companies hurried to copy its design. With its transverse engine, front-wheel drive and surprising outside dimensions, the Mini was from then on the benchmark in the modern compact car segment.

In Britain only a limited number of compact cars were available. Consumers had the choice of the BMW Isetta for £390, the whimsical Heinkel Cabin Roller for £399 and the Goggomobil T300 for £495. All three models were exports from West Germany, had rear-wheel drive and an extremely loud motorcycle engine. Their price and running costs alone were more affordable than for 'conventional' cars. Quite a number of Britons opted for an Austin A35, a Morris Minor 1000 or a Standard Eight. These could at least reach 100 km/h and ran on around 9 litres per 100 km. With such benefits, even hard suspension, limited space for four and questionable road-holding were seen as acceptable.

So if we want to assess the role of the MINI, we must remember that criteria such as these were not insignificant. One of the first two models of the Mini, the Austin Seven, cost a grand £496. For a Citroën 2CV, on the other hand, one had to splash out a bit more and come up with £564, while a Fiat 600 came in at £613 and a VW Beetle £617. At such an affordable price, the Mini fit right in with Sir Alec Issigonis's original dream of a Mini parked outside every worker's home. Whether in terms of forecast or ideal, the car was unable to meet the sales expectations of the 1960s.

The situation in Germany was no different: 'It was a time to put away the past, which had always stood in the way of progress, and to bring about the realization of consumer society, which had taken shape in the 1920s, had fuelled the hopes and dreams of the masses in the 1930s and had now, on the ruins of the old world, truly become a reality. Apartment, car, travel – the desires of the 1920s had turned into

hatte, in den Dreißigerjahren in den Erwartungshorizont der Massen gerückt war, und jetzt, auf den Trümmern der alten Welt, endlich zur Verwirklichung anstand. Wohnung, Auto, Reisen, aus den Wünschen der Zwanzigerjahre waren in der Vorkriegszeit Versprechungen geworden, die sich nun als Bedürfnisse geltend machten, um dann in den Sechzigerjahren zu Ansprüchen umzuschlagen …«

In Deutschland hatte die Motorisierung nach dem Zweiten Weltkrieg nur schleppend begonnen. Bis Mitte der Fünfzigerjahre beherrschten Motorräder und Motorroller das Straßenbild der Großstädte. Während der Motorradbestand langsam zurückging, wuchs die Zahl zugelassener Automobile stetig an. Wer bisher bei Wind und Wetter mit dem Motorrad zur Arbeit gefahren war, stieg nun verstärkt auf ein kleines Auto um – Hauptsache, ein festes Dach über dem Kopf.

Ein typisches Fahrzeug des Übergangs war das von Glas gebaute Goggomobil. 1955 kam es auf den Markt und durfte dank seines laut knatternden 250 ccm-Motors mit dem bisherigen Motorradführerschein gefahren werden. Im selben Jahr brachte BMW als Lizenz die Isetta heraus, ein Rollermobil mit runden Formen und nur 2,28 Meter Länge, auch als »Motocoupé« bezeichnet. Hoch im Kurs standen auch kleine, wendige Automobile wie der Fiat 500, ganze 2,97 Meter lang, der 1957 vorgestellt wurde. Im Folgejahr, 1958, präsentierte der damals weltgrößte Motorradhersteller NSU mit dem Prinz seinen ersten Nachkriegswagen, angetrieben durch einen 20-PS-Heckmotor.

Mit dem Modell 700 brachte BMW 1959 ein viersitziges Sportcoupé heraus. Der Kleinwagen mit dem modernen Design und einem 30 PS starken Motor wurde zum Erfolgsmodell und bereitete den Weg für den dauerhaften Aufstieg der Marke BMW im Verlauf der Sechzigerjahre. Der Bestseller auf dem deutschen Markt aber war und blieb der VW Käfer: 1955 wurde die Millionengrenze überschritten, 1959 sogar die Drei-Millionen-Marke erreicht. Der MINI aber tat sich auf dem deutschen Markt schwer. Das lag im Wesentlichen am Kaufpreis, den der Importeur auf 5780 DM setzte und sich damit gegen starke Konkurrenten wie den Volkswagen für 4600 DM und den nagelneuen BMW 700 Sport für 5650 DM durchsetzen musste.

promises in the pre-war era, thereafter asserting themselves as needs before becoming entitlements in the 1960s…'.

In Germany after World War II the uptake of the motor car had been hesitant. Until the mid-1950s, motorcycles and motor scooters had dominated cities. While motorcycles slowly reduced in numbers, the proportion of registered cars went up. Many of those who had previously driven to work in all weathers on their motorcycles now increasingly changed over to a compact car – the main thing was to have a fixed roof over your head.

The Goggomobil by Glas was one such transitional vehicle. Released on the market in 1955, its rattling 250 cc engine meant it could be driven by anyone with a motorcycle licence. That same year, BMW launched the Isetta, a round-shaped roller with a length of only 2.28 m, also called the 'Motocoupé'. Small, manoeuvrable cars were very popular, such as the Fiat 500 at a length of 2.97 m, which was released in 1957. The following year, 1958, the then world's biggest motorbike manufacturer NSU launched its first post-war car, the Prinz, which was powered by a 20 bhp rear engine.

In 1959, BMW presented the 700, a four-seater coupé. With its modern design and a 30 bhp engine this compact car was a success and paved the way for the on-going advancement of the BMW brand throughout the 1960s. But the VW Beetle was, and was set to remain, *the* bestseller on the German market: hitting the threshold of one million cars sold in 1955, it even reached three million by 1959. The Mini, however, struggled in its first years on the German market. This was largely due to its sales price, set by the importer at 5,780 DM, which meant that it was faced with major competitors such as a Volkswagen at 4,600 DM and a brand new BMW 700 Sport for 5,650 DM.

Das grandiose Fahrzeugkonzept des Mini beschränkte sich nicht auf den Hatch, also die zweitürige Limousine mit abgeschrägtem Heck, sondern wurde in den folgenden vierzig Jahren auf eine Vielzahl faszinierender und funktionstüchtiger Karosserievarianten angewandt. Die brillanten Lösungen, die der Konstrukteur Issigonis bei der Entwicklung des familientauglichen Zweitürers gefunden hatte, reichten für weit mehr als einen einzigen Autotyp und ließen sich auf weitere Modelle übertragen. Nur wenige Fahrzeugkonzepte der Automobilgeschichte haben vergleichbar lange Zeitspannen überdauert und keines mit einer ähnlich großen Vielfalt bedacht.

Im Herbst 1961 wurde das Standardprogramm vorübergehend durch Modelle mit geringfügig größerem Komfort erweitert. Als neue Ausführungen stellte BMC den Austin Super Seven und Morris Super Mini-Minor vor. Doch beschränkten sich die Neuerungen auf ein überarbeitetes Armaturenbrett, das die separaten Instrumente Tacho, Öldruckanzeiger und Kühlwasserthermometer aufwies. Die dezente Überarbeitung des Urmodells zählt zu den Konstanten der MINI-Historie seit 1959.

DIE KLASSISCHEN MODELL-VARIANTEN / THE CLASSIC MODELS

The fantastic concept of the Mini was not limited just to the basic, two-door saloon car with a truncated rear. The following 40 years saw a multitude of fascinating and highly functional chassis types. The brilliant solutions that designer Issigonis had come up with in the development of the family-friendly two-door car could be used for much more than a single vehicle type and could be translated into other models. In the history of the automobile only a few vehicle concepts have survived as long, and none has been favoured with as much variety.

In the autumn of 1961, a slightly more comfortable version was temporarily added to the standard programme. As new versions BMC introduced the Austin Super Seven and Morris Super Mini-Minor. However new features were limited to a re-vamped dashboard, with separate speedometer, oil-pressure and water-temperature gauges. Since 1959 subtle revisions of the original model have taken place throughout the history of the MINI.

Mini Van

Im Juni 1960 – der Hatch war noch kein Jahr auf dem Markt – erweiterte BMC die Palette um einen schlichten, aber vielseitig einsetzbaren Lieferwagen, den Van. Es handelte sich hierbei um einen zweisitzigen Kastenwagen, dessen Plattform 25 Zentimeter länger ausfiel als das Standardmodell. Auch war der Radstand um 10 Zentimeter länger und das Dach geringfügig höher als beim Vorgänger. Zwei handliche Flügeltüren am Heck des fensterlosen und geschlossenen Laderaums erleichterten das Be- und Entladen.

Bis zur Hinterkante der Tür, der sogenannten B-Säule, waren Nutzfahrzeug und Standardlimousine, also der Hatch, baugleich. Der sich anschließende kastenförmige Aufbau fügte sich mit seinen Proportionen harmonisch an. Vor allem bei Handwerkern und Transportunternehmen erfreute sich der Mini Van großer Beliebtheit. So war es nicht verwunderlich, dass auch die britische Post den wendigen Lieferwagen für den Zustelldienst orderte. Selbst die britische Polizei und Automobilclubs im Lande zählten zu den Großabnehmern des Van.
/
The Mini had been on the market for less than a year when in 1960 BMC added a simpler, but more versatile delivery vehicle to its range – the Van. This was a two-seater panel van whose platform was 25 cm longer than that of the standard model. Its wheelbase was also around 10 cm longer and the roof slightly higher than that of its predecessor. Two handy hinged doors at the rear of the panelled-in, enclosed cargo space facilitated loading and unloading.

Up to the back edge of the front doors, the so-called B-pillar, both the commercial vehicle and standard saloon car, the Hatch, were of identical construction. The proportions of the box-shaped platform that was added created a harmonious fit. The Mini Van was especially popular among tradesmen and transport companies, so it came as no surprise when the Royal Mail ordered this nippy delivery vehicle for its postal service. The British police and automobile clubs were also among bulk buyers of the Van.

Mini Wildgoose

Sie ist unbestritten eines der ungewöhnlichsten Exemplare der Modellpalette von MINI, ein skurriles Wohnmobil mit dem abenteuerlichen Namen »Wilde Gans«. Zwischen 1963 und 1968 hatte die Firma Wildgoose Ltd. in Worthing, Sussex, rund sechzig kleine Wohnmobile mit ausstellbarem Dach auf Basis des Mini Van gebaut. Heute haben sich davon noch etwa zehn Exemplare weltweit erhalten, nur wenige dürften in fahrtauglichem Zustand sein. Die Wildgoose Brent Super V.E.B. mit der Seriennummer 18 wurde 1965 gebaut und galt als Topmodell. Für den Antrieb sorgte ein Vierzylinder mit 850 ccm und 34 PS, der ein Tempo von 80 Stundenkilometern gut meistert, sich aber in seiner Bauart gegen die angegebene Spitzengeschwindigkeit von 116 Stundenkilometern ausspricht. Man denke auch an die winzigen 10-Zoll-Räder, die das Ungetüm mit stattlichen 622 Kilogramm Leergewicht tragen mussten.

Die Wildgoose ist in erster Linie ein gepflegter Reisebegleiter und eine mobile Urlaubsunterkunft. Die Prospekte warben mit »Überall hinfahren! Überall bleiben!« und sprachen vornehmlich »Pensionärsehepaare« an, die »von den Zwängen des Berufslebens befreit eine sorgenlose Freizeit genießen« wollten. 998 Britische Pfund – der doppelte Kaufpreis eines Mini Van – musste ihnen dieses Erlebnis schon wert sein. Das Wohnmobil zeichnete sich durch Wirtschaftlichkeit und Komfort aus und bildete ein prägnantes Beispiel für den »Creative Use of Space«, der MINI eigen ist. Kaum 4 Meter lang, bietet es Sitz- und Schlafmöglichkeit für vier Personen. Nach dem Abendessen wird der Tisch mithilfe einer mehrteiligen Stabkonstruktion zur Unterlage des Bettes, das zwei Personen angenehme Ruhe verspricht. Der Clou ist ein elektrisch ausfahrbares Teleskopdach, das die Stehhöhe auf luxuriöse 1,80 Meter erweitert. Die Aluminiumwände des Schlafabteils klappen nach unten und werden mit Spannhebeln arretiert. Hier sind Liegen für zwei weitere Personen vorgesehen. Die Verwandlung in ein Schlafzimmer ist bei geübtem Handgriff binnen weniger Minuten möglich. Zum besonderen Komfort gehören außerdem ein zweiflammiger Gaskocher, eine Spüle, Vorhänge sowie reichlich Stauraum. Für eine unverwechselbare Note sorgt eine Zweifarben-Lackierung der Außenhaut in Türkis und Elfenbein.
/
This is undoubtedly one of the most unusual models of the MINI, a whimsical camper van with the spirited name of Wildgoose. Between 1963 and 1968, Wildgoose Ltd. a company based in Worthing, Sussex, had constructed around 60 small camper vans with an extendable roof based on the Mini Van. Today, around ten still survive scattered across the world, but only a few of them are likely to be still drivable.

The Wildgoose Brent Super V.E.B. with serial number 18 was built in 1965 and sold as the top model. With its four-cylinder drive of 850 cc and 34 bhp, it can easily reach 80 km/h, although its build makes the stated top speed of 116 km/h a little tricky – just imagine the tiny 10-inch tyres that had to carry the impressive 622 kg of unladen weight.

Above all the Wildgoose is a sleek travel companion and a mobile holiday home. The brochures advertised it with the slogan 'Go anywhere! Stop anywhere!', primarily addressing 'retired couples' who 'being freed from business, desire the carefree life'. At £998 – twice the price of a Mini Van – the Wildgoose must have been well worth it to them. The camper van was characterised by economy and comfort and provided a good example of the 'Creative Use of Space' unique to the Mini. Barely 4 m long it provides seating and beds for four. After dinner, a multi-part rod construction can be used to turn the table into the base of the bed, which can accommodate two people comfortably. But its chief attraction is an electrically extendable telescopic roof, which increases the standing room to a luxurious 1.80 m. The aluminium walls of the sleeping compartment can be folded down and fastened with clamps to create sleeping quarters for another two people. With practice, the transformation into a bedroom takes just a few minutes. The camper van's unique luxuries also include a two-burner gas cooker, sink unit, curtains and lots of storage space. Its distinctive character comes from its two-tone paint finish in turquoise and ivory.

Mini Ice Cream Van > S./p. 160

Eine außergewöhnliche Nutzung fand der Mini Van in Gestalt des Ice Cream Van. Ursprünglicher Besitzer des Wagens war eine Ice Cream Company namens Trufelli mit Sitz in Sheffield, South Yorkshire. Im Lauf der Jahre wechselte der Wagen mehrfach den Besitzer – zeitweise stand er in Diensten von Tony's Ices – und wurde dabei in unterschiedlichsten Farben lackiert. Der heutige Besitzer, der den Van vor mehr als zehn Jahren erwarb, erinnert sich an einen sehr schlechten Erhaltungszustand. Unter anderem mussten zahlreiche Farbschichten abgetragen werden, ehe die ursprüngliche Lackierung mit dem Logo von Trufelli wieder zum Vorschein kam.
/
One extraordinary use of the Mini Van was in the shape of an ice cream van. The van was originally owned by an ice cream company called Trufelli, headquartered in Sheffield, South Yorkshire. Over the years, the car changed owner several times – for a while, it was used by Tony's Ices – and was painted in a variety of different colours. The current owner, who bought the van more than ten years ago, recalls that it was in a very poor condition. For instance, several layers of paint had to be removed before the original paintwork with the Trufelli logo was revealed.

Mini Outspan Orange

Zu den originellen Sonderkarosserien zählt der Outspan Orange, eine fahrbare Zitrusfrucht, der man die Erbanlagen eines Mini auf den ersten Blick kaum ansieht.

In den späten Sechzigerjahren kam die südafrikanische Kooperative Citrus Exchange and Capespan auf den Gedanken, in außergewöhnliche Werbung für die Marke Outspan Orange zu investieren, um in Großbritannien und auf dem europäischen Kontinent ihre Orangen erfolgreich absetzen zu können. Folglich gab das Unternehmen bei Brian Waite Engineering in Sussex sechs »Promotional vehicles« in Form von Orangen in Auftrag. Die Basis bildete ein Mini mit Automatikgetriebe und einem Motor mit einem Fassungsvermögen von 998 ccm Hubraum. Design und Konstruktion des Aufbaus waren besonders anspruchsvoll. Indem man den Radabstand auf 1,20 Meter verkürzte, wurde das Chassis dem Grundriss einer kreisrunden Orange angepasst. Um eine ausreichende Gewichtsverteilung und damit eine einigermaßen akzeptable Straßenlage zu erreichen, wurde am hinteren Hilfsrahmen ein 90 Kilogramm schwerer Betonblock angebracht.

Die einer Orangenhaut nachempfundene äußere Hülle besteht aus gepresstem Fiberglas. Für die Fensterscheiben nutzte man orange abgetöntes Plexiglas, eine unscheinbare Tür am Heck bildet den Eingang. Die Innenausstattung ist komplett in orangefarbenem Vinyl gehalten.

Im Laufe der Siebzigerjahre rekrutierte das Unternehmen in Südafrika zahlreiche junge Damen, um sie in Europa für Werbezwecke einzusetzen. Sie trugen orangefarbene Uniformen sowie Perücken und zogen in Begleitung der Orangen-Mini durch mehrere Länder, um auf Volksfesten und vor Geschäften für Outspan Orange zu werben. Die Kampagne war vor allem deshalb notwendig geworden, weil das Apartheidregime in Pretoria weltweit für Kritik sorgte und einen Boykott südafrikanischer Produkte auslöste.
/
The Outspan Orange, a citrus fruit on wheels, is one of the original special chassis, whose relationship to a Mini might not be immediately obvious. In the late 1960s, the South African cooperative Citrus Exchange and Capespan had the idea of investing in unusual advertisements for the Outspan Orange brand in order to sell their oranges successfully in Great Britain and Europe. So the company ordered six orange-shaped 'promotional vehicles' from Brian Waite Engineering in Sussex. The automatic Mini with a 998 cc engine formed the basis of the vehicle. Design and construction of the platform were particularly challenging. By shortening the wheelbase to 1.20 m, the chassis was customized to fit the plan of a circular orange. In order to distribute the load sufficiently and thus maintain relatively acceptable road-holding, a 90 kg concrete block was attached to the rear auxiliary chassis.

The outside shell made to resemble orange peel is in moulded fibreglass. Orange-tinted perspex was used for the windows, and an inconspicuous door at the rear formed the entrance. The interior consists entirely of orange-coloured vinyl.

During the 1970s, the company recruited several young ladies in South Africa to advertise its products in Europe. They wore orange uniforms and wigs and travelled through several countries with the Orange Mini to advertise Outspan Orange to businesses and at fairs. The campaign had become particularly necessary because the apartheid regime in Pretoria had been criticized all over the world, causing a boycott of South African products.

Mini Kombi / Estate: Austin Seven Countryman, Morris Mini Traveller, Pick-up
> S. / p. 164

Der zunehmende Erfolg des Mini, vor allem im Segment der kleinen Nutzfahrzeuge, führte im September 1960 zur Präsentation des Kombimodells und im Januar 1961 des Pick-up. Beide besaßen die Außenmaße des Van, doch faszinierte das Kombimodell durch seine Rundumverglasung. Überaus praktische Elemente waren die zwei Hecktüren und eine umklappbare Rückbank. Der Laderaum mit insgesamt 1,20 Meter Länge und 85 Zentimetern Breite war immens. Das Ladevolumen betrug folglich 1000 Liter, die Zuladung 317 Kilogramm. Wie sich schon beim Hatch bewährt hatte, wurde auch der Kombi in zwei Varianten – als Morris Mini Traveller und gleichzeitig als Austin Seven Countryman – vorgestellt, wobei die Bezeichnungen unmissverständlich die jeweiligen Zielgruppen ansprachen. Die damalige Werbung setzte die unterschiedlichen Käufer und Einsatzmöglichkeiten auf plakative Weise in Szene.

Die Gestaltung des Traveller mit Blendstreben aus Esche auf den hinteren Seitenflächen und am Heck fiel ins Auge. Zwar war der komplette Aufbau des Wagens aus Stahl, doch kam BMC mit diesem etwas antiquierten Ornament dem damals in Großbritannien vorherrschenden Geschmack entgegen. Die hölzernen Streben waren eher eine nostalgische Zutat, erinnerten sie doch an den legendären Woody, wie der Morris Minor Traveller aus den Fünfzigerjahren mit Spitznamen hieß. Dieser »Volkswagen Englands« war seit 1948 gebaut worden und für seine Zuverlässigkeit und Genügsamkeit berühmt.

Wie der Mini trägt auch das geniale Fahrzeugkonzept des Morris Minor Traveller die Handschrift des Designers und Entwicklers Alec Issigonis.

Extrem vielseitig war auch das kleinste Modell aus der Riege der Lastesel, der Pick-up. Er war im Gewerbe wie auch in der Landwirtschaft ein geschätztes Nutzfahrzeug. Zahlreiche Sonderbauten wurden für ihn entwickelt.
/
The growing success of the Mini, especially in the small commercial vehicles sector, led to the launch of the estate model in September 1960 and of the Pick-up in January 1961. With the same dimensions as the Van, the estate car could boast all-round windows. Its highly practical elements included two rear doors and a folding back seat. Measuring 1.20 m long and 85 cm wide the cargo space was enormous. Consequently the cargo volume amounted to 1,000 litres, with a load capacity of 317 kg. Following the success of the Mini saloon, the estate car was also launched in two versions – the Morris Mini Traveller and at the same time the Austin Seven Countryman –

with the two names unmistakably addressing the respective target groups. Contemporary ads were aimed at the different buyers with posters showing how each car could be used.

With its ash-wood trim on the rear sides and tail the design of the Traveller was eye-catching. Although the entire car was made of steel, BMC accommodated British taste of the time with this somewhat antiquated decoration. The wooden trim was a rather nostalgic addition, reminiscent of the legendary 'Woody', the nickname given to the 1950s' Morris Minor Traveller. This 'English Volkswagen' had been built after 1948 and was famous for its reliability and economy.

Much like the Mini, the inspired concept of the Morris Minor Traveller is also recognisably the brainchild of designer and developer Alec Issigonis. Even the smallest model of this series of 'workhorses', the Pick-up, was also extremely versatile. It was a utility vehicle that was very popular with both tradesmen and in agriculture, and many customized models were developed.

Riley Elf, Wolseley Hornet
> S. / p. 168

Eine Sonderrolle kommt den Marken Wolseley und Riley zu. Wolseley war 1927 von Morris gekauft worden, 1938 folgte Riley. Gemeinsam mit Morris bildeten sie seit 1938 die Nuffield-Gruppe und gehörten somit dem BMC-Imperium an. Beide Marken genossen seit den Dreißigerjahren einen hervorragenden Ruf und knüpften mit qualitativ hochwertigen, komfortablen Limousinen in den Fünfzigerjahren daran an, wobei Riley eher die sportlich ausgerichtete Klientel bediente, während Wolseley die luxuriösere Ausstattung anbot.

Riley und Wolseley bauten auf der Urversion des Mini von 1959 auf, mit einigen Veränderungen. Auf den ersten Blick wird man die Verwandtschaft zum originalen Mini suchen müssen: Zu sehr präsentieren sich die Modelle, welche Riley Elf (Elf = Elfe) und Wolseley Hornet (Hornet = Hornisse) benannt wurden, als Luxuswagen der unteren Mittelklasse. Die beiden Modelle heben sich durch ein eigenes Design, vor allem durch einen Respekt einflößenden Kühler, einen sichtbar ausgebildeten Kofferraum, verstärkte Stoßstangen und Schwalbenschwänzchen-Kotflügel am Heck ab.

Besonders deutlich wird der Unterschied beim Interior Design: Die Fahrgastzelle war zwar baugleich, doch bot der Riley Elf mehr Luxus: Das Armaturenbrett aus gediegenem Wurzelholz nahm die gesamte Breite ein und integrierte zwei abschließbare Handschuhfächer. Der Schaltknauf war verchromt, die Türverkleidung überarbeitet. In Fachmagazinen wurde dieses Modell scherzhaft gern als »kleiner Rolls-Royce« bezeichnet. Nachdem der Konzern British Motor Corporation 1968 durch Fusion zu British Leyland geworden war, führten strenge Rationalisierungsmaßnahmen 1969 zur Einstellung beider Modelle.

The brands Wolseley and Riley occupy a very special place. Morris bought Wolseley in 1927, followed by Riley in 1938, and since then together with Morris, they formed the Nuffield Group, making them part of the BMC empire after the merger in 1952. Since the 1930s, both brands had enjoyed excellent reputations, continuing to produce high-quality, comfortable saloon cars into the 1950s, with Riley's customers focusing primarily on a sporty clientele while Wolseley offered luxury designs.

Both Riley and Wolseley built upon the original 1959 version of the Mini, with several changes. At first glance, the relationship to the original Mini is barely noticeable. By contrast, the models Riley Elf and Wolseley Hornet were marketed as small luxury cars for the lower middle classes. The two models stand out due to their unique designs, especially the stylised front grille, visibly extended boot, reinforced bumpers and dovetail mudguards at the rear.

The interior design makes this difference even more apparent: the basic passenger compartment may have been identical, but the Riley Elf's fixtures and fittings were more luxurious. The solid burr-walnut dashboard extended the entire width of the interior, integrating two lockable glove compartments. The gear knob was chrome, and the door panels had been redesigned. Car magazines liked to joke that this was a 'baby Rolls-Royce'. After the British Motor Corporation had merged in 1968 to become British Leyland Motor Corporation, tough rationalisation measures in 1969 led to the discontinuation of both models.

This car had been a response to a request by the British armed forces for an extremely light off-road vehicle. The experience of previous military engagements had shown that troop supply from the air would become increasingly important, and the provision of light off-road vehicles could provide a crucial advantage. The total weight was to remain below the 600 kg threshold so that it would be possible to transport the car by helicopter or to drop it off at its destination by parachute. In late 1962, Alec Issigonis was able to present the development of the compact off-road vehicle, essentially modelled on a Mini Countryman. The twin-engined Moke powered the front and back axles separately. Its relationship to the Mini was limited to the wheelbase and some of its technology. Its platform build, however, was an entirely new design.

The robust construction, small dimensions and low weight of the Moke might have made it possible to drop such a vehicle by parachute, but its insufficient ground clearance and lack of four-wheel drive meant that the Moke was unusable as a military off-road vehicle. As a result, BMC focused on how the car could be used in a civilian context. The first orders came from agriculture and forestry. But the small off-roader found its niche among leisure users and jet-setters. It became a genuinely cult car of the Swinging Sixties

and was used at the beach, in golf clubs or in marinas. The Moke may have offered few comforts, but it promised pure driving fun.

The Moke scored its biggest sales successes in the sun-kissed regions of California and Australia, and in 1966, production was moved to Australia. In 1968, the new owner, British Leyland Motor Corporation, decreed the immediate discontinuation of the fun car. In 1980, production was moved once more, this time to Portugal.

Mini Clubman > S. / p. 184

Im Oktober 1969 ergänzte der Mini Clubman das Programm. Es handelte sich hierbei um die erste Neuentwicklung eines Mini-Modells unter Führung von British Leyland.

Mit 11 Zentimetern übertraf er deutlich die Fahrzeuglänge des klassischen Mini. Äußerlich unterscheidet sich der Mini Clubman durch eine veränderte Frontpartie. Die modernisierte Ausführung mit der komprimierten Fassung von Kühlergrill und Scheinwerfern sowie der formalen Integration der vorderen Kotflügel stieß bei Puristen auf Ablehnung. Die spürbar verlängerte Front widersprach zwar dem ursprünglichen Credo eines möglichst kurzen Autos, doch war

Mini Moke > S. / p. 176

Im August 1964 stellte BMC eine weitere Variante des Mini vor, die ursprünglich für den militärischen Einsatz konzipiert worden war: den Mini Moke. Seine Karosserie bestand aus einer Bodenwanne mit breiten, kastenförmigen Schwellern plus Motorhaube und Windschutzscheibe. Gegen Niederschläge schützte allenfalls ein aufklappbares Verdeck.

Vorausgegangen war eine Anfrage der Britischen Streitkräfte, die nach einem extrem leichten Geländewagen verlangten. Die Erfahrungen der vergangenen Kriegseinsätze hatten gezeigt, dass der Truppenversorgung aus der Luft eine zunehmende Bedeutung zukommen würde, bei der das Abwerfen leichter Geländefahrzeuge mitunter entscheidend sein könnte. Das Gesamtgewicht sollte die 600-Kilogramm-Marke nicht überschreiten, sodass der Wagen ohne Probleme von einem Hubschrauber getragen oder mit Hilfe eines Fallschirms über seinem Bestimmungsort abgeworfen werden konnte. Ende 1962 legte Alec Issigonis das Konzept eines kompakten Geländewagens vor, der im Wesentlichen auf einem Mini Countryman basierte. Bei dem zweimotorigen Antrieb wurde die Leistung getrennt auf die Vorder- und Hinterachse übertragen. Die Verwandtschaft zum Mini beschränkte sich auf den Radstand und Teile der Technik. Der Plattformaufbau hingegen war eine komplette Neukonstruktion.

Die robuste Bauweise, die geringen Abmessungen und das niedrige Gewicht des Moke (engl. = Esel) erfüllten zwar die Voraussetzung, ein solches Fahrzeug per Fallschirm landen zu lassen, doch eine zu geringe Bodenfreiheit und der fehlende Allrad-Antrieb machten den Moke für einen militärischen Geländeeinsatz unbrauchbar. BMC konzentrierte sich daraufhin auf seine zivile Einsatzmöglichkeit. Erste Aufträge kamen aus der Land- und Forstwirtschaft. Doch Karriere machte der kleine Offroader im Bereich von Freizeit und Jetset. Er avancierte zu einem regelrechten Kultauto der Swinging Sixties und kam am Strand, in Golfclubs oder Yachthäfen zum Einsatz. Der Moke bot zwar keinerlei Annehmlichkeiten, verhieß jedoch puren Fahrspaß.

Größte Absatzerfolge erzielte der Moke in den sonnenverwöhnten Regionen Kaliforniens und in Australien. 1966 wurde seine Produktion nach Australien verlegt. 1968 verfügte der neue Eigentümer British Leyland das sofortige Aus des Fun Car. 1980 wurde die Fertigung noch einmal nach Portugal verlagert.
/
In August 1964, BMC launched another version of the Mini, originally designed for use by the military – the Mini Moke. Its body consisted of a base unit with wide, box-section pontoons plus bonnet and windscreen. A fold-down hood was the only protection against the elements.

sie aufgrund verschärfter Crashvorgaben erforderlich geworden. Auch bot sie die Möglichkeit, zukünftig größere Motoren einbauen zu können.

Serienmäßig war der Mini Clubman mit einem 38 PS starken 1-Liter-Motor ausgestattet. Ihm stellte man mit dem Mini 1275 GT ein karosseriegleiches Pendant zur Seite, dessen 1,3-Liter-Motor eine Leistung von 59 PS erbrachte. Im Gegenzug jedoch wurde der Mini Cooper aus dem Programm genommen. Ebenfalls neu im Angebot war der Mini Clubman Estate, der die bisherige Kombi-Version ablöste. Auch hier kam man etwaigen Kundenwünschen mit ausgeprägtem Hang nach Holz entgegen, allerdings reduziert auf einen handbreiten Streifen aus Holzfurnier, der entlang der Flanken und am Heck verlief, eingefasst in Chromleisten.
/
In October 1969 the Mini Clubman was added to the product range. This was the first newly developed model of a Mini under the management of British Leyland.

It was 11 cm longer than the classic Mini, and externally, the Mini Clubman boasted an altered front end. The updated finish with its more compact radiator grille and headlights and the formal integration of the front wings was rejected by purists. The markedly longer front end may have been contrary to the original aim of a car

that was as short as possible, but more stringent anti-crash measures had made this necessary. Moreover this presented the opportunity of being able to install bigger engines in the future.

The Mini Clubman was equipped with a 38 bhp 1-litre engine as standard. Its companion model, the Mini 1275 GT used the same body and could boast a 1.3-litre engine and 59 bhp. In exchange, however, the Mini Cooper was discontinued. The new range also included the Mini Clubman Estate, which replaced the previous estate car. This too accommodated any possible customer taste with a marked tendency towards the use of wood, albeit reduced to a 15 cm wood-effect strip running down the sides and the tail, with chrome edging.

Mini Cabrio/Convertible
> S./p. 188

1991 wurde schließlich die letzte Variante des klassischen Mini vorgestellt: das Mini Cabrio. Gut drei Jahrzehnte mussten vergehen, bis die Marke das Fahren mit offenem Verdeck ermöglichte.

Als einziges Modell hat das Cabrio seinen Ursprung nicht in England, sondern in Deutschland. Ein engagierter Mini-Händler aus dem badischen Kappelrodeck bei Achern hatte – wie schon manch anderer Tuner-Betrieb vor ihm – das Dach des Mini abgeschnitten und die Karosserie in ein äußerst ansehnliches Cabriolet verwandelt. Im Unterschied zu früheren Versuchen überzeugte dieser Karosserieumbau mit verstärktem Schweller und der Integration eines Querträgers durch besondere Qualität. Und erreichte damit die Verantwortlichen der Rover Group, die inzwischen für die Marke MINI verantwortlich war und Konstruktion beziehungsweise Fabrikation kaufte.

Das Dach war handwerklich ausgereift, zugfest, wasserdicht und dick gefüttert. Ab 1993 wurden die klassischen Mini Cabrio im MINI-Werk in Longbridge gebaut. Bis August 1996 entstanden dort insgesamt 1081 Fahrzeuge, die heute zu den begehrten Raritäten zählen. Cabrios lagen zu Beginn der Neunzigerjahre sehr im Trend. Gefragt waren vor allem die Ausführungen in Nigthfire Red mit rotem oder Caribbean Blue mit grauem Verdeck.
/
In 1991, the last version of the classic Mini was launched – the Mini Convertible. It had taken a good three decades before the brand came up with an open-top driving experience. The Convertible is the only model to have originated not in England, but in Germany. Like many businesses before him a dedicated Mini dealer from Kappelrodeck near Achern in Baden had cut off the roof of the Mini and turned the chassis into an impressive cabriolet. But in contrast to previous attempts, this chassis conversion was particularly high quality with its reinforced rocker panel and integrated cross-beam. This struck a chord with management at the Rover Group, which were by this time responsible for the MINI brand, and so they bought up the design and production.

The roof was skilfully engineered, draught-proof, waterproof and thickly padded. From 1993, the classic Mini Convertible was built at the MINI Works in Longbridge. By August 1996, a total of 1,081 vehicles had been constructed there, and today they are considered sought-after rarities. In the early 1990s, Convertibles were extremely fashionable, with the Nightfire Red edition featuring a red hood, or Caribbean Blue with a grey hood both selling well.

Mini Shorty > S./p. 172

Der Fantasie experimentierfreudiger Karosseriewerkstätten als auch privater Alleskönner sind kaum Grenzen gesetzt, wenn es um die individuelle Neuinterpretation eines bestehenden MINI geht. In der klassischen Mini-Szene, die sich mindestens einmal im Jahr auf den legendären International MINI Meetings trifft, werden die besonderen Kreationen diskutiert, bestaunt und begutachtet. Zu den Meistern seines Fachs zählt Kevin Palmer, der seit Jahrzehnten gemeinsam mit seinem Vater eine MINI-Werkstatt vor den Toren Londons betreibt. Sein Lieblingsmodell ist der in grellem Pink leuchtende Mini Shorty. Das ursprüngliche Modell stammt aus dem Jahr 1963. 2002 – zum fünfzigjährigen Jubiläum des väterlichen Betriebs – wurde der Mini in eine Kurzversion, den sogenannten Mini Shorty, verwandelt: Die Karosserie ist von der Front bis zum Armaturenbrett identisch mit herkömmlichen Modellen, die Rückfront aber wurde von der üblichen Gesamtlänge von 3,05 Metern auf 2,20 Meter zusammengestaucht: Da blieb kein Platz mehr für Kofferraum und Seitentüren, weshalb Kevin heute schwungvoll in den Fahrersitz springt. Der gekürzte klassische Mini erntet, wo immer er auftritt, Applaus und Begeisterung. Für freudiges Gelächter sorgt auch ein hinten sichtbar aufgesetzter, kleiner Spezialtank aus Edelstahl, gefüllt mit Lachgas, das dem Sprit beigemischt wird. Dem umgearbeiteten klassischen Modell setzte Kevin 2007 mit einer ebenso überzeugenden Überarbeitung des neuen MINI noch eins drauf – auch hier ist ihm kollektive Bewunderung sicher.
/
As far as individual re-interpretations of the MINI are concerned, there are hardly any limits to the imagination of adventurous body shops and private hobbyists. Classic Mini enthusiasts, who meet at least once a year at the legendary International MINI Meetings, discuss, marvel at and assess these special creations. One such expert is Kevin Palmer who for decades has been running his MINI workshop just outside London with his father. His favourite model is the outrageously pink Mini Shorty. The original model dates from 1963. In 2002 – for the 50th anniversary of his father's business – the Mini was converted into a truncated version, the so-called Mini Shorty. From the front end to the dashboard the body is identical to the regular model, but the back part of the front end was reduced to 2.20 m from its original 3.05 m. This left no space for a boot or side doors, which is why Kevin has to jump energetically into the driver's seat. This abbreviated classic Mini encounters cheers and enthusiasm wherever it goes. Another source of great amusement is the small, add-on stainless steel tank, which can be attached to the back and filled with laughing gas to be mixed with the fuel. In 2007, Kevin topped his customized classic model with an equally impressive revision of the new MINI – earning him general admiration once again.

Erstes deutschlandweites MINI-Treffen auf der Loreley / First nationwide MINI meeting on the Lorelei in Germany, 2005

Die größte »Power« in der Welt des MINI steckt nicht unter der Motor-haube eines MINI Cooper S oder eines John Cooper Works, sondern geht von der internationalen MINI-Community aus. Seit 1959 finden Millionen von Fans zueinander, die ihre Begeisterung für den kleinen Wagen »made in Britain« teilen. Und es werden immer mehr. Allein das legendäre International MINI Meeting, in Fachkreisen kurz IMM genannt, zieht seit seinem Ursprung 1978 Tausende von Besuchern aus aller Herren Länder an. Die gigantische Veranstaltung, die jährlich stattfindet und die größte ihrer Art in Europa ist, wird durch MINI-Clubs privat organisiert und bietet jeweils an drei Sommertagen jede Menge Unterhaltung, Fahrspaß sowie einen faszinierenden Teilemarkt und ist gleichermaßen Infoplattform als auch Kontaktbörse.

Doch was wäre dieses Ereignis ohne eine gewachsene und mitt-lerweile in aller Welt etablierte MINI-Clubszene? Mit besonderer Lei-denschaft pflegt diese die Tradition der Marke und bringt Menschen zusammen, die ihren MINI nicht nur dazu benutzen, von A nach B zu kommen. Für sie ist MINI eine Weltanschauung, eine permanente In-spiration. Kein Wunder, dass die internationale MINI-Community in der automobilen Welt einzigartig ist. Um teilzunehmen, werden oftmals keine Mühen gescheut: Den längsten Anfahrtsweg nahm ein Besucher aus Moskau auf sich, der in fünf Tagesetappen rund 3660 Kilometer zurücklegte. Und das alles nur, um dabei zu sein.

Schon zu Beginn der Sechzigerjahre wuchs in der britischen Hei-mat des MINI eine aktive und bunt gemischte Fangemeinde heran. Hinsichtlich Technik und Zubehör tauschte man Wissen und Erfahrun-gen aus, unternahm gemeinsame Ausfahrten und kleine Treffen soge-nannter Stammtische, die sich bereits vor dem Zeitalter des Internets überregional zu vernetzen wussten.

Auch in Deutschland hat sich eine lebendige Clubszene etabliert, die auf mehr als 6000 Mitglieder in rund hundert MINI-Communities und über fünfzig MINI-Clubs angewachsen ist. Ihre charmanten und fantasievollen Aktivitäten machen sie zu wahren Markenbotschaftern.

MINI-Liebhaber treffen sich aber nicht nur auf der Straße, sondern auch online. Vor allem das Internet bietet die Möglichkeit zur Kommu-

COMMUNITY

The biggest 'power' in the world of the MINI is not what can be found under the bonnet of a MINI Cooper S or a John Cooper Works, it is in fact the international MINI community. Since 1959, millions of fans have got together to share their enthusiasm for this compact 'made in Britain' car, and their numbers are increasing. The legendary Interna-tional MINI Meeting, professionally known as IMM, has on its own at-tracted thousands of visitors from all over the world since it was first held in 1978. This enormous event, which takes place annually and is the biggest of its kind in Europe, is organised privately by MINI clubs and offers plenty of entertainment, driving fun and a fascinating parts market, as well as an information platform and a networking opportu-nity over three days in the summer.

But what would this event be without a growing and by now glob-ally established MINI club scene? The latter upholds the tradition of the brand with a particular passion, and brings together people who use their MINI for more than just getting from A to B. For members, MINI is a way of life and a permanent inspiration. No wonder that the international MINI community is unique in the world of cars, and to take part its members often make a huge effort: the longest journey was made by one attendee from Moscow who drove a total of around 3,660 km in five days just to be there.

Right from the early 1960s, an active and diverse community of fans was formed in Britain, the home of the MINI. They swapped knowl-edge and experiences about technology and accessories, went on group trips and organised small regular meetings, which developed ways of networking across the regions long before the age of the Inter-net.

Germany too saw the establishment of a lively club scene, which has now grown to include more than 6,000 members in around 100 MINI communities and more than 50 MINI clubs. Their appealing and imaginative activities have made them into true brand ambassadors.

But lovers of the MINI don't just meet on the road – they also meet online. The Internet in particular offers ways of communicating and in-teracting across national boundaries and continents. MINI has been

nikation und Interaktion über Ländergrenzen und Kontinente hinweg. Seit 2008 ist MINI im Bereich Social Media aktiv. Für besonders kreative MINI-Fahrer ist MINISPACE.com mehr als eine Internetplattform. Sie führt Menschen, Events und Projekte aus aller Welt zusammen, initiiert und begleitet Wettbewerbe, Veranstaltungen sowie von unterschiedlichen Communities getragene Aktionen. Aufbauend auf dem Grundgedanken des Konstrukteurs Sir Alec Issigonis, der mit dem MINI einst das maximale Auto auf minimaler Verkehrsfläche schuf, propagiert MINISPACE.com den »Creative Use of Space« und überträgt diese Idee auf urbane Projekte. Im öffentlichen Raum als auch im privaten Umfeld wird durch Kreativität und Interaktion auf kleinster Fläche ein Höchstmaß an Spaß, Lifestyle und Freiraum für neue Ideen geschaffen. Inzwischen trifft sich eine weltweite Community auf MINISPACE.com und hat Anteil an verschiedenen Projekten. Jedes Mitglied kann sie mit seinem Input bereichern oder Anregungen für eigene Aktivitäten gewinnen. Die Marke MINI ist ungemein aktiv. So finden sich Millionen von Fans auf den Social Media Channels wie Facebook, Twitter, Instagram, Google+ und bei YouTube.

active in the realm of social media since 2008. For highly creative MINI drivers, MINISPACE.com is more than just an online platform. It brings together people, events and projects from across the world, launching and supporting competitions and events as well as activities in different communities. Based on designer Sir Alec Issigonis's original idea of the Mini as the car with maximum interior space on a minimum footprint, MINISPACE.com promotes the 'Creative Use of Space' and transfers these ideas to urban projects. Publicly and privately, creativity and interaction on a minimum footprint create a maximum of fun, lifestyle and space for new ideas. Today, MINISPACE.com plays host to a global community that takes part in many different projects. Each member can provide his or her input to these projects, or ask for suggestions concerning their own activities. The MINI brand is amazingly active. It has millions of fans on social media channels such as Facebook, Twitter, Instagram, Google+ and on YouTube.

MINI-Fans treffen sich auf der ganzen Welt / MINI fans meet up from all over the world

Einen MINI zu fahren, ist neben dem praktischen Aspekt der Fortbewegung oftmals auch Ausdruck des individuellen Stils. Aus diesem Grund bietet MINI Möglichkeiten an, den Wagen nach eigenen Wünschen zu gestalten – eine große Palette an Außenlackierungen, Innenraumfarben, Sitzbezügen und Dekor-Varianten steht hierfür zur Auswahl. Unter den meistgewählten Lacktönen rangiert an erster Stelle Chili Red, ein kraftvoll leuchtendes Rot, gefolgt von Pepper White, Pure Silver, Black und Hyper Blue.

Erste Bestandteile einer maßgeschneiderten Konfiguration werden bereits ab Werk angeboten und lassen sich durch das Original MINI-Zubehörprogramm ergänzen. Zu diesem zählen klassische Rennsporttechnik in Form von John-Cooper-Works-Komponenten, komfortable und nützliche Details wie iPod-Schnittstelle, Dachträger aus hochfestem Aluminiumprofil oder Zusatzscheinwerfer. Da die Produktions- und Logistikprozesse flexibel auf die Variantenvielfalt ausgerichtet sind, ist es äußerst unwahrscheinlich, dass im Lauf eines Jahres zwei identische Fahrzeuge das Band im Werk Oxford verlassen.

Wer seinem MINI ein neues Erscheinungsbild verleihen möchte, kann mittels Zubehör für Abwechslung sorgen: Außenspiegelkappen, Seitenblinkereinfassungen, Radventilkappen und Türgriffe im Union-Jack- sowie im Checkered-Flag-Design, ein Tankverschluss in Chromausführung, Sport- oder Bonnet-Stripes und unterschiedlich geformte Leichtmetallfelgen bereichern das Sortiment. Ganz oben auf der Wunschliste vieler Kunden steht die Farbwahl für das Dach. Zur Verfügung stehen eine Vielzahl an Dekorvarianten, die spannende Kontraste zur Farbe der Karosserie setzen oder Lieblingsmotive wie den Union Jack abbilden.

Schon der klassische Mini der ersten Jahrzehnte bediente den Wunsch nach stärkerer Individualisierung: Legendär sind ausgefallene Sonderanfertigungen berühmter Stars und Persönlichkeiten. Denn der revolutionäre Kleinwagen regte schon in den Sechzigerjahren die Fantasie von Schauspielern, Modedesignern, Musikern und Hochadel an. Sonderlackierungen und luxuriöse Ausstattungsoptionen standen bei ihnen hoch im Kurs. So gab Peter Sellers in den Siebziger-

INDIVIDUALITÄT / INDIVIDUALITY

In addition to the practical aspect of transport, driving a MINI can often be an expression of individual style as well. This is why MINI offers opportunities of customizing the car based on customer requests – a wide range of exterior and interior colours, seat covers and décor alternatives is available. Chilli Red (a bright red) is one of the colours most frequently chosen, followed by Pepper White, Pure Silver, Black and Hyper Blue.

The first components of a tailored configuration are already offered ex-works and can be supplemented with original MINI accessories. These include traditional racing technology in the shape of John Cooper Works components, and comfortable and useful details such as an iPod connection, high-strength aluminium-profile roof-rack or additional headlights. Since production and logistics processes provide a flexible focus on variety, it is highly unlikely that two identical vehicles would leave the plant at Oxford during the course of a year.

Anyone wishing to give their MINI a new appearance can spruce it up with accessories: outside mirror caps, indicator mounting, wheel valve caps and door knobs in the designs of the Union Flag and Chequered Flag, a chrome petrol cap, sporty or bonnet stripes and aluminium rims in different shapes contribute to the range. At the top of many customers' wish list is the colour of the roof. Many options are available, providing an exciting contrast with the colour of the car's body or illustrating favourite motifs such as the Union Flag.

The classic Mini of the first decades already had a lot to offer in the way of individual expression, and unusual one-off designs for famous stars and personalities have become legendary. In the 1960s the revolutionary compact car was already feeding the imaginations of actors, fashion designers, musicians and the aristocracy. Special colours and luxurious interiors were very popular. In the 1970s, for example, Peter Sellers ordered several spectacular custom-made versions of the Mini in one go. A member of the Brunei royal family received an individually tailored and tuned Mini in a flower-power design in 1988. Mini launched its first special editions in the late-1970s, with sporty models focusing on a youthful look or the hint of upper-class elegance. Famous

jahren gleich mehrere spektakuläre Mini-Sonderanfertigungen in Auftrag. Ein Mitglied des Herrscherhauses von Brunei erhielt 1988 einen nach individuellen Wünschen ausgestatteten und getunten Mini im Flower-Power-Design. Mitte der Siebzigerjahre bot Mini erstmals Sondereditionen an, Modelle mit sportlicher Note, betont jugendlichem Erscheinungsbild oder angedeuteter Noblesse. Auch bekannte Londoner Stadtteile und Straßenzüge wie Piccadilly, Chelsea, Knightsbridge oder Park Lane standen häufig Pate. Den Anfang machte der auffällig progressiv wirkende Mini Limited Edition 1000 des Jahrgangs 1976, 1982 folgte ihm mit dem Mini Mayfair erstmals ein Modell der Spitzenklasse mit betont exklusiver Ausstattung auf den Markt.

areas and roads in London were also often an inspiration, such as Piccadilly, Chelsea, Knightsbridge or Park Lane. The progressive-looking Mini Limited Edition 1000 led the way in 1976, followed by the Mini Mayfair in 1982, the first top-class model with an exclusive interior.

3

MOTORSPORT

In den Fünfzigerjahren kaufte die Cooper Car Company bei der BMC Teile ein, woraufhin sich zwischen beiden Unternehmen ein intensiver Kontakt entwickelte. John Cooper, der mit Alec Issigonis befreundet war, hatte schon während der Entstehung der ersten Prototypen das sportliche Potenzial des neuen Kleinwagens erkannt. Noch vor der eigentlichen Premiere des Mini machte er sich mit großem Elan ans Tunen. Zwar hatte Issigonis, der primär das Ziel eines alltagstauglichen Autos für jedermann verfolgte, diesbezüglich Vorbehalte, doch durfte Cooper mit der Genehmigung von BMC-Konzernchef George Harriman eine Kleinserie von 1000 Stück produzieren. Den Motor dieser Mini Cooper vergrößerte er auf 1,0 Liter Hubraum und erhöhte dadurch die Leistung auf 55 PS bei 6000 l/min. Seine Variante des Mini erreichte nun eine maximale Geschwindigkeit von 140 km/h und zeigte sich damit deutlich agiler als der 116 km/h schnelle Mini des Standardmodells.

Die Reaktionen auf dieses Auto, das im September 1961 auf den Markt kam, waren euphorisch und riefen nach noch mehr Leistung. In Zusammenarbeit mit Issigonis und Daniel Richmond von Downton Engineering konnte Cooper den Hubraum des Mini-Motors der A-Serie auf 1071 ccm steigern und blieb damit knapp unter dem Limit der angepeilten Rennsportklasse von 1100 ccm. Der Motor erreichte beeindruckende Drehzahlen und leistete sensationelle 70 PS. Und damit nicht genug: Die Übersetzung des Getriebes wurde an das sportliche Potenzial angepasst, größere Scheibenbremsen an den Rädern der Vorderachse gewährleisteten entsprechende Verzögerungswerte und waren servounterstützt. Bald bildete der Mini Cooper S nicht nur auf der Straße eine Ausnahmeerscheinung. Der Klassensieg des Finnen Rauno Aaltonen bei der Rallye Monte Carlo im Jahr 1963 markierte nur den Auftakt für eine beispiellose Erfolgsserie im Motorsport. Sie gipfelte in den drei Gesamtsiegen bei der Rallye Monte Carlo in den Jahren 1964, 1965 und 1967.

Von 1961 bis 1971 wurde der Mini Cooper als Serienfahrzeug angeboten. Der Name Cooper war nun weltweit bekannt und stand für Leidenschaft und Fahrspaß. Auch Besitzer regulärer Mini-Fahrzeuge

JOHN COOPER

In the 1950s, the Cooper Car Company bought engines for its F3 racing cars from the British Motor Corporation (BMC), bringing the two companies into close contact. John Cooper, who was friends with Alec Issigonis, had realized the sporty potential of the new Mini right from the very first prototypes, before its launch in 1959. But before the actual première of the Mini, he started to think about tuning the car, taking it to the Italian Grand Prix so he could understand the car better. Although Issigonis had initial doubts, since his primary goal was to produce a practical car for everyday use, Cooper completed his first prototype in 1960, and as a result, was commissioned to produce a small batch of 1,000 units with the approval of George Harriman, Chairman and Managing Director of the BMC. He enlarged the Mini's A-Series engine to 1,000 cc, thus increasing power to 55 bhp at 6,000 rpm. His Mini variant then had a top speed of 140 km/h, decidedly faster than the 116 km/h of the standard Mini model.

The market reacted euphorically when the car was launched in September 1961 with some calling for even more performance. In cooperation with Issigonis and Daniel Richmond of Downton Engineering, Cooper again increased the size of the Mini's A-Series engine to 1,071 cc, just slightly below the limit of the racing class target of 1,100 cc. The engine achieved impressive revs and had a much improved power output of 70 bhp. But that was not enough. The gear ratios were adapted to the sporty potential and larger front disc brakes provided the necessary deceleration. Soon the Mini Cooper S became an unusual sight, not only on the roads. The class victory by the Finn Rauno Aaltonen in the Monte Carlo Rally in 1963 marked the start of an unmatched series of successes in motor racing. It culminated in three overall victories in the Monte Carlo rally in the years 1964, 1965 and 1967.

From 1961 to 1971, the Mini Cooper and Cooper S were offered as production vehicles. By now, the Cooper name had become famous all over the world and stood for passion and driving pleasure. Owners of standard Mini cars were also impressed by the tuning kits bearing the John Cooper Works brand, introduced later in the 1980s. Despite its

1923

Am 17. Juli wurde John Cooper in Surbiton/Surrey geboren. Seit früher Kindheit faszinierten ihn Bau und Entwicklung von Autos für den Einsatz im Motorsport.

/

John Cooper was born in Surbiton, Surrey on 17 July. From early childhood, he was fascinated by building and designing cars for use in motor racing.

1947

Gemeinsam mit seinem Vater Charles Cooper gründete er die Cooper Car Company, die in Surbiton in einer kleinen Werkstatt Rennwagen für die Formel 2, Formel 3 und später

Formel 1 baute. John Cooper bestritt bereits mit zwölf Jahren sein erstes Rennen.

/

Together with his father Charles Cooper he founded the Cooper Car Company to build racing cars for Formula 2, Formula 3 and later Formula 1 in a small garage in Surbiton. John Cooper ran his first race at the age of 12.

1955

Die Coopers waren für ihre technischen Neuerungen bekannt. So führten sie in der Formel 1 den Mittelmotor ein, der nicht mehr wie üblich vor, sondern hinter dem Piloten platziert wurde.

/

The Coopers were known for their technical breakthroughs: for example they introduced rear-engine cars to Formula 1. The engine was positioned behind the driver's seat instead of in the usual place in front.

1958

John Coopers Team feierte die ersten Erfolge in der Formel 1.

/

John Cooper's team celebrated its first successes in Formula 1.

1959/60

Rennwagen »made by Cooper« holten den Konstrukteurs- und Fahrertitel in der Königsklasse. Jack Brabham wurde in einem Cooper Formel-1-Weltmeister. Zu den bekanntesten Cooper-

Piloten zählen außer ihm Stirling Moss, Bruce McLaren, Niki Lauda und Jochen Rindt.

John Cooper und sein Team bedienten sich des klassischen Mini, der von seiner Konzeption her dafür ausgelegt war, praktisch, sparsam und geräumig zu sein. Als Motoreningenieur trieb Cooper die Entwicklung des Mini jedoch maßgeblich an und brachte ihn förmlich auf Touren. Mit dem legendären Mini Cooper wurde der Kleinwagen zum Inbegriff des englischen Automobils der Sechzigerjahre. John Cooper legte damit den Grundstein für John Cooper Works.

/

Racing cars 'made by Cooper' won the constructors' and

drivers' titles in the première class. Jack Brabham became Formula 1 world champion in a Cooper. Other famous Cooper drivers included Stirling Moss, Bruce McLaren, Niki Lauda and Jochen Rindt.

John Cooper and his team used the classic Mini with its design concept based on practicality, fuel economy and spaciousness. As an engine specialist, Cooper was instrumental in pushing forward the development of the Mini's power unit, literally bringing it up to full revs. With the legendary Mini Cooper the small car was the epitome of the British giant-killing performance car of the 1960s. John Cooper then laid the founda-

tion stone of the John Cooper Works.

2000

John Cooper starb am 24. Dezember im Alter von 77 Jahren.

/

John Cooper died on 24 December at the age of 77.

---

zeigten sich von den Tuning-Kits der Marke John Cooper Works begeistert. Trotz seiner Popularität stellte BMC 1971 die Produktion des Mini Cooper nach einer Stückzahl von mehr als 150 000 gefertigten Autos ein. John Cooper und sein Sohn Michael hielten den Namen der Marke am Leben. Um die wachsende Nachfrage zu bedienen, produzierten sie in den Achtzigerjahren Tuning-Kits und Zubehör.

Im Jahr 1990 schließlich wurde Mini Cooper wieder ins Modellprogramm aufgenommen und erfreut seither MINI-Fans in aller Welt. John Cooper Works bildet heute unter dem Dach von MINI eine eigenständige Marke. Sie profitiert von traditionsreichem Know-how im Motorrennsport und bildet den Inbegriff für extremen Fahrspaß. Inzwischen tragen mehrere MINI-Modelle das besondere Markenlogo, das herausragende Sportlichkeit verkörpert, darunter der MINI John Cooper Works, der MINI John Cooper Works Clubman und das MINI John Cooper Works Cabrio. Angetrieben werden sie von kraftvollen 155-KW-Motoren. Zugleich bietet ein Sortiment verschiedener Ausstattungen Raum für Individualität, höchste Sicherheitsstandards und einen unverwechselbaren MINI-Style. Bei einer Fahrt auf der Landstraße oder professionellen Einsätzen im Motorsport bieten die Modelle von MINI John Cooper Works umfassende Technikpakete. Das Zubehör reicht von Optionen im Bereich Aerodynamik und einem Sportfahrwerk mit rot lackierten Federn, gelochten Bremsscheiben, einer Domstrebe für den Motorraum bis hin zu Spiegelkappen aus Karbon oder Seitenblinkereinfassungen mit sportiver Gitterstruktur. Im Innenraum sorgen Zusatzinstrumente, Dekorleisten, Handbremshebel und Sportschalthebel im Karbon-Look für sportives Ambiente.

popularity BMC stopped production of the Mini Cooper in 1971 after it has manufactured more than 150,000 cars. John Cooper and his son Michael kept the name of the brand alive. To meet growing demand, they produced tuning kits and accessories in the 1980s.

Finally in 1990, the Mini Cooper was once again included in the model range, to the delight of Mini fans all over the world. Today, John Cooper Works forms an independent brand under the MINI umbrella. It benefits from the rich tradition of motor racing know-how and is the epitome of exceptional driving pleasure. In the meantime, several MINI models bear the special emblem which is the embodiment of superior sportiness, including the MINI John Cooper Works, the MINI John Cooper Works Clubman and the MINI John Cooper Works Convertible. They are driven by powerful 155 kW engines. At the same time, a wide range of different equipment levels leaves room for individuality, very high safety standards and a distinctive MINI style. Whether it is for driving on rural roads or in professional motor sports, the models from MINI John Cooper Works offer an extensive range of tuning packs. Accessories range from aerodynamic kits and sport suspension with red coil springs, perforated brake discs, a strut brace for the engine compartment through to mirror caps made of carbon fibre or side direction indicator surrounds with sporty grid pattern. In the interior, additional instruments, décor strips, handbrake handles and sports gearshift levers in carbon create a sporty look.

Folgende Seiten / Following pages

Rallye Monte Carlo / Monte Carlo Rally, 1964

Die Erfolge der Marke im internationalen Rallyesport reichen von den Einsätzen in der FIA World Rally Championship bis zu Teilnahmen an Rallye Raid-Veranstaltungen wie der weltberühmten Rallye Dakar. Der Rallye-Sport zeichnet sich durch eine Besonderheit aus: Alle Piloten haben hier denselben Gegner: die Uhr. Anders als bei Rennen auf der Rundstrecke kämpfen die Fahrer in dieser Disziplin nicht im direkten Duell gegeneinander, sondern müssen auf einem festgelegten Abschnitt – der sogenannten Wertungsprüfung – die schnellste Zeit im Teilnehmerfeld erzielen. Unterstützt wird jeder Fahrer von einem Kopiloten, der zuvor den detaillierten Verlauf der Piste notiert hat und seinem Kollegen am Steuer präzise Anweisungen geben kann. Zu jeder Rallye gehören traditionell auch Verbindungsetappen zwischen den einzelnen Start- und Zielpunkten, die auf öffentlichen Straßen und Wegen zurückgelegt werden. Für jeden dieser Abschnitte gibt es eine Minimal- und eine Maximalzeit. Verfehlt ein Fahrer diese Zeiten, werden Strafen ausgesprochen.

Die Rallye Monte Carlo gilt weltweit als die prestigeträchtigste Rallye überhaupt. Das seit 1911 ausgetragene Rennen stellt in unterschiedlichen Wertungsprüfungen die Leistungsfähigkeit der Autos und Fahrer unter fast allen erdenklichen Fahrbedingungen auf die Probe. Die meisten Etappen müssen in den französischen Seealpen zurückgelegt werden – gefürchtet ist dabei der letzte Abschnitt am 1607 Meter hoch gelegenen Col de Turini. Dreimal, 1964, 1965 und 1967, konnte Mini diese Rallye für sich entscheiden und dabei die übermächtige Konkurrenz mit ihren V8-Motoren hinter sich lassen.

2011 und 2012 kehrte MINI in den Rallyesport zurück und schickte bei der FIA World Rally Championship (WRC) den neu konstruierten MINI John Cooper Works WRC, der auf Basis des MINI Countryman von Prodrive entwickelt wurde, erfolgreich ins Rennen. Der wohl größte Erfolg gelang Dani Sordo (ES) im Januar 2012, als er beim Comeback von MINI bei der Rallye Monte Carlo den zweiten Platz erringen konnte. Während diese Veranstaltungen traditionell auf abgesperrten Straßen oder eigens präparierten Pisten stattfinden und die Gesamtdistanz oft nicht mehr als 500 Kilometer beträgt, werden die sogenannten Rallye-

RALLEYS UND RUNDSTRECKEN-RENNEN / RALLY RACING AND CIRCUIT RACING

The brand's successes in international rally racing range from events in the FIA World Rally Championship to entries in Rally Raid events and the world-famous Dakar Rally. Rally racing is well known for one particular feature: all drivers compete against the same opponent – the clock. Unlike racing on a circuit, drivers in this discipline are not duelling directly with each other. Their aim is to achieve the fastest time across the field of competitors in fixed sections known as special stages. Every driver is supported by a co-driver who takes detailed notes of the track in advance so he can give his colleague at the wheel precise instructions. Traditionally every rally has connecting legs run on public roads and tracks between each of the starting and finishing points. There are minimum and maximum times for each of these stages. If a driver fails to keep to these times, penalties are awarded.

The Monte Carlo Rally is regarded as the most prestigious rally in the world. The race, which has taken place since 1911, tests the performance of both car and driver under almost every conceivable driving condition in a number of different special stages. Most of the stages must be driven in the French Maritime Alps – and the dreaded last stage passes over the Col de Turini at an altitude of 1,607 m. The Mini won this rally three times, in 1964, 1965 and 1967, leaving behind a field of superior competitors, some sporting V8 engines.

In 2011 and 2012, the MINI returned to rally racing and successfully sent the newly designed MINI John Cooper Works WRC, which was developed by Prodrive based on the MINI Countryman, to race in the FIA World Rally Championship (WRC). Probably the greatest success was achieved by Dani Sordo (ES) in January 2012 when he gained second place in the Monte Carlo Rally to celebrate the MINI's come-back. While these events traditionally take place on roads closed to traffic or on specially prepared tracks and the total distance is often no more than 500 km, the so-called Rally Raid long-distance rallies cover many thousands of kilometres in open terrain. Rally Raid drivers themselves are particularly tough, often driving 500 km in one day.

The most famous long-distance rally is the Dakar Rally, which has been staged since 1979. Due to the unstable political situation from

Raid-Langstreckenrallyes über Tausende von Kilometern im offenen Gelände ausgetragen. Rallye-Raid-Piloten sind besonders hart im Nehmen, oft absolvieren sie 500 Kilometer an einem Tag.

Die berühmteste Marathonrallye ist die Rallye Dakar, die seit 1979 ausgetragen wird. Aufgrund der unsicheren politischen Verhältnisse wurde dieses härteste Offroad-Rennen der Welt ab 2009 nicht mehr auf dem afrikanischen Kontinent ausgetragen, sondern in Südamerika. Die neue Strecke umfasst 8400 Kilometer, führt über Wüstensand, Monsterdünen und schneebedeckte Andenpässe und passiert dabei mehrere Staaten des Subkontinents. 2011 ging erstmals auch ein MINI an den Start. Schon ein Jahr später folgte die Sensation: Der Franzose Stéphane Peterhansel fuhr in einem MINI ALL4 Racing, der auf Basis eines John Cooper Works Countryman entwickelt worden war, als Erster durchs Ziel, und war begeistert: »Dieses Auto war definitiv ausschlaggebend für den Sieg. Es gab nicht ein einziges technisches Problem. Wir hatten keine Panne. Das habe ich noch nie mit einem Auto erlebt.« Damit nicht genug: Alle fünf eingesetzten MINI-Rennwagen schafften es unter die ersten zehn Plätze. Nach 2012 konnten die MINI-Piloten auch 2013 und 2014 den Gesamtsieg einfahren und ein Stück Motorsportgeschichte schreiben.

Im Rundstrecken-Rennsport hat sich die Marke MINI im Lauf ihrer fast sechzigjährigen Geschichte auf vielen namhaften Rennstrecken rund um den Globus etabliert. So legte sich der Kleinwagen auf dem Parcours in Brands Hatch mit größeren und leistungsstärkeren Konkurrenten an. Ob bei den legendären Bergrennen auf den französischen Mont Ventoux, auf dem Nürburgring, in Snetterton – der Mini ging an vielen Orten an den Start und feierte unzählige Siege und Meisterschaften. Er wurde nicht nur zu einem prägenden Rennfahrzeug der Sechzigerjahre, sondern trug auch viel dazu bei, den Tourenwagensport beim Publikum wieder attraktiv zu machen. Nicht selten gelang es MINI alias David, Goliath davonzufahren, und wenn einmal nicht den Sieg, so doch die Sympathie der Zuschauer zu gewinnen. Große Fahrernamen verbinden sich mit dem Engagement von MINI auf der Rundstrecke, so John Fitzpatrick, John Handley oder Ken Tyrrell. James

Stéphane Peterhansel mit Beifahrer / with co-driver Jean-Paul Cottret, 2013

2009 onwards, the world's toughest off-road race has no longer taken place on the African continent but in South America. The new route covers 8,400 km, crossing desert sand, monster dunes and snow-covered Andes passes, and travelling through several countries on the sub-continent. A MINI first entered the race in 2011. A year later came the sensational news that the Frenchman Stéphane Peterhansel was the first to cross the finishing line in a MINI ALL4 Racing which is based on a John Cooper Works Countryman. 'This car was definitely the decisive factor in winning the victory. There was not a single technical hitch. We had no breakdowns. I've never experienced that with a car before.' But that was not all: all five of the MINI cars entered in the race finished among the first ten places. After 2012, MINI drivers also managed to win the overall victory in 2013 and 2014, thus writing a new chapter in motor sports history.

In the course of its almost 60-year history, the MINI brand has gained a reputation on many famous racing circuits all around the world. For instance, the small car was pitted against larger and more powerful competitors on the circuit in Brands Hatch. Whether it is in the legendary mountain race on Mont Ventoux in France, on the Nürburgring or in Snetterton, the Mini has entered races in many places and has won innumerable victories and championships. Not only was it a pioneering racing car in the 1960s, it has also contributed greatly to making touring-car racing attractive to the public once more. Time and again the MINI might look like David but manages to leave Goliath standing. If it does fail to win, at least it gains the sympathy of the spectators. Some great drivers have been associated with MINI's commit-

Folgende Seiten / Following pages

MINI Challenge, 2004

Mini-Rennen im Crystal Palace / Mini racing at Crystal Palace, 1966

Rallye Monte Carlo / Monte Carlo Rally, 1964

Rallye Dakar / Dakar Rally, 2014

Hunt, Jochen Rindt und Niki Lauda bestritten ihre ersten Rennen mit einem Mini.

Ein besonderes Event im Motorsport ist die als MINI Challenge bekannt gewordene Rennsport-Serie. Sie wird seit 2004 ausgetragen und hat sich inzwischen zu einer der erfolgreichsten und populärsten Clubsportserien der Welt entwickelt. Ihre Attraktivität verdankt die MINI Challenge, in Deutschland seit 2012 als MINI Trophy bezeichnet, einem innovativen Konzept. »Motorsport trifft Lifestyle« ist eine Mischung, die bei Fahrern genauso wie bei Fans und Zuschauern ankommt. Zur besonderen Atmosphäre trägt die Besetzung des Teilnehmerfelds bei. Neben Routiniers und Nachwuchstalenten aus dem Motorsport gehen auch Prominente aus Showbusiness und Sport an den Start. Das wohl bedeutendste Kennzeichen des Reglements ist die absolute Chancengleichheit: Alle Piloten starten mit technisch identischen Fahrzeugen ins Rennen. Jeder Mini John Cooper Works Challenge, der hier teilnimmt, ist ein speziell für den Rennsport umgebauter MINI. Wichtige Komponenten sind versiegelt, um unerlaubte Veränderungen auszuschließen. Aufgrund der identischen technischen Voraussetzungen für alle Teams und Teilnehmer sind packende Duelle während der Rennen garantiert. Das souveräne Handling des MINI und die anspruchsvolle Sicherheitsausstattung der Rennfahrzeuge machen die MINI Challenge vor allem für Rennsporteinsteiger und ambitionierte Amateurpiloten interessant.

Auf der Rennstrecke und im Fahrerlager ist Action angesagt. Tausende von Zuschauern, Fahrern und Promis wollen MINI-Challenge-Rennwochenenden nicht missen. Wo sonst kommt man in den Fahrerlagern den Piloten und ihren Teams so nahe und kann echtes Racing-Feeling hautnah miterleben? In jedem Jahr werden an acht Rennwochenenden 16 Läufe ausgetragen. Sie finden im Rahmen hochkarätiger Motorsportevents in Deutschland, Belgien, Spanien und der Schweiz, in Australien, Brasilien und Argentinien statt.

ment to circuit racing, such as John Fitzpatrick, John Handley and Ken Tyrrell. James Hunt, Jochen Rindt and Niki Lauda drove their first races in a Mini.

A special event in motor sports is the racing sport series known as the MINI Challenge. It has taken place since 2004 and since then it has grown to become one of the most successful and popular club sport series in the world. The MINI Challenge owes its popularity to an innovative concept and has been known in Germany as the MINI Trophy since 2012. 'Motor sports meets lifestyle' is a mixture that is just as popular with drivers as it is with fans and spectators. The qualifications of the entrants contribute to the special atmosphere at the events. Alongside seasoned and young up-and-coming drivers from motor sports, celebrities from show business and sport also line up at the start. The most significant feature of the rules is the absolute equality of opportunity. All drivers start the race with technically identical cars. Every Mini John Cooper Works Challenge taking part is a MINI that has been specially tuned for racing. Key components are sealed to prevent unauthorized modifications. The identical technical conditions for all teams and entrants guarantee thrilling duels during the races. The MINI's superior handling and the sophisticated safety equipment on the racing cars make the MINI Challenge interesting, not only for beginners in motor racing but also for ambitious amateur drivers.

There is no shortage of action both on the racing circuit and in the drivers' paddock. Thousands of spectators, drivers and celebrities would never miss the MINI Challenge race at weekends. Where else can you be so close to drivers and their teams in the drivers' paddocks, and experience close-up that genuine racing feeling? Every year 16 races take place on eight race weekends, scheduled during top-class motor racing events in Germany, Belgium, Spain, Switzerland, Australia, Brazil and Argentina.

Der Motorsport der Marke MINI verzeichnet eine Vielzahl erfolgreicher und couragierter Rennfahrer, allen voran die »Drei Musketiere« Rauno Aaltonen, Timo Mäkinen und Paddy Hopkirk, die in den Sechzigerjahren die Rallyewelt im Sturmlauf eroberten und bei der Rallye Monte Carlo im Mini Cooper S triumphierten.

/

The MINI brand in motor sports has known many successful and courageous racing drivers, in particular the 'Three Musketeers' Rauno Aaltonen, Timo Mäkinen and Paddy Hopkirk, who took the rally world by storm in the 1960s and triumphed in the Monte Carlo Rally in the Mini Cooper S.

Rauno Aaltonen
Am 4. Juni 1938 wurde Rauno Aaltonen im finnischen Turku geboren. Der wohl weltweit berühmteste Rallyefahrer erlangte Bekanntheit unter dem Spitznamen »der Rallye-Professor«, weil er sich stets akribisch mit der Technik des Rallyefahrens auseinandersetzte.

Aaltonen fuhr 1944 im Alter von sechs Jahren erstmals auf dem Hof seiner Eltern Auto und Motorrad. Mit 13 Jahren absolvierte er sein erstes Motorbootrennen, mit 16 wurde er Speedway-Motorradweltmeister und gleichzeitig skandinavischer Motorbootmeister.

Von 1962 bis 1968 war Aaltonen Rallye-Werksfahrer für MINI. 1963 fuhr Aaltonen im Mini Cooper S als Klassensieger und Drittplatzierter des Gesamtklassements ins Ziel und sorgte bereits hier für eine Sensation.

1966 wurde der Mini Hattrick zum Triumph der Rennsaison. Den Fahrern Aaltonen, Tommi Mäkinen und Peter Hopkirk gelang ein Geniestreich, als sie unmittelbar nacheinander, als Erster, Zweiter und Dritter, durchs Ziel fuhren. Doch wurden sie disqualifiziert, weil die Ab-blendvorrichtung der Hauptscheinwerfer ihrer Rennwagen nicht der Homologation entsprochen hatte.

Im Mini Cooper S holte sich Rauno Aaltonen 1967 den dritten Monte-Sieg und wurde damit zu einem der erfolgreichsten Rallye-Fahrer aller Zeiten.

Seit 1977 war Aaltonen unter anderem als Chefinstruktor für das BMW Fahrertraining auf Eis und Schnee verantwortlich, das in seiner Heimat Finnland ausgetragen wird.

/

Rauno Aaltonen was born in Turku, Finland on 4 June 1938. Probably the best-known rally driver, he came to fame with the nickname 'The Rally Professor' for his ongoing and meticulous study of the techniques of rally driving.

In 1944 at the age of six Aaltonen first drove a car and a motorcycle belonging to his parents in their backyard. At the age of 13, he drove his first power-boat race and at 16 he became speedway world motorcycle champion and Scandinavian power-boat champion simultaneously.

From 1962 to 1968, Aaltonen was rally works driver for Mini. 1963 When Aaltonen crossed the finishing line in the Mini Cooper S as class winner in 1963, he created a sensation, coming third in the overall rankings.

In 1966 the Mini Hattrick triumphed in the racing season. Drivers Aaltonen, Tommi Mäkinen and Peter Hopkirk created a real stroke of genius by crossing the finishing line one after the other in first, second and third places. But they were disqualified because the dipped-beam mechanism on the main headlights of their racing cars failed to comply with the homologation requirements.

In 1967, Rauno Aaltonen won his third Monte Carlo Rally in the Mini Cooper S, thus becoming the most successful rally driver of all time.

Since 1977, Aaltonen has been responsible as Chief Instructor for BMW driver training on snow and ice, leading courses in his home country of Finland.

Paddy Hopkirk
Am 14. April 1933 wurde Paddy Hopkirk im nordirischen Belfast geboren.

Im Rallye-Winter 1963/64 feierte er mit einem Mini Cooper S seinen größten sportlichen Erfolg: In einem spektakulären Rennen holten Hopkirk und sein Kopilot Henry Liddon den ersten Platz im Gesamtklassement der Rallye Monte Carlo. Mit diesem Geniestreich war aus dem kleinen Sprinter eine Legende im Motorsport geworden. Die Startnummer 37 und das Autokennzeichen 33 EJB sind seither im Rennsport von einer gewissen Symbolkraft. Paddy Hopkirk erinnerte sich: »Der Mini war ein sehr fortschrittliches Auto, obwohl er nur ein kleiner Familienwagen war. Sein Frontantrieb und der vorne quer eingebaute Motor waren sehr vorteilhaft, genauso wie die Tatsache, dass das Auto klein war und die Straßen kurvig. Sie waren außerdem ziemlich schmal, was meiner Meinung nach auch ein Vorteil für uns war. Und wir hatten viel Glück, dass die Autos in Ordnung waren, dass alles zur richtigen Zeit passierte und zum richtigen Zeitpunkt zusammenkam.«

Michael Cooper und Hopkirk gaben gemeinsam wichtige Hinweise bei der Konstruktion des neuen MINI.

/

Paddy Hopkirk was born in Belfast, Northern Ireland, on 14 April 1933. In 1963/64 he enjoyed his greatest sports success in a Mini Cooper S during the winter rally season. In a spectacular race, Hopkirk and his co-driver Henry Liddon took first place in the overall rankings of the Monte Carlo Rally. With this sensation the small-time sprinter became a legend in motor sports. Since that time, start number 37 and license plate 33 EJB have become iconic in motor racing. Paddy Hopkirk remembers: 'The Mini was a very advanced car although it was only a small family saloon. Its front-wheel drive and the transverse mounted front engine had many advantages, just as much as the fact that the car was small and the roads were curvy. They were also very narrow, which in my opinion was also an advantage for us. And we were very lucky that the cars ran perfectly and that everything happened at the right time and at the right moment.' Together Michael Cooper and Hopkirk offered important tips for the design of the new MINI.

Timo Mäkinen
1965 wiederholte der Finne Timo Mäkinen, der im März 1938 geboren war, mit seinem Kopiloten Paul Easter den Monte-Triumph. Er war in der Zeit von 1962 und 1968 Mitglied im Mini Rallye Werksteam. Als einziger Fahrer meisterte er Tausende von Kilometern ohne einen einzigen Strafpunkt – und das trotz widrigster Witterungsbedingungen. Nur 35 von 237 gestarteten Autos erreichten bei dieser Rallye das Ziel, darunter drei Mini Cooper S. Mit dem Mini wurden auch weiterhin Gewinne eingefahren, darunter drei Gesamtsiege in Serie bei der 1000-Seen-Rallye 1965, 1966 und 1967.

/

In 1965 the Finn Timo Mäkinen, who was born in March 1938, repeated the Monte Carlo triumph together with his co-driver Paul Easter. He was a member of the Mini rally works team from 1962 to 1968. He was the only driver to complete the thousand kilometres without losing a single penalty point, despite extremely adverse weather conditions. Only 35 of the 237 cars that started the rally reached the finishing line, among them three Mini Cooper S cars. The Mini also won three other races, including three overall wins in succession at the 1000 Lakes rally in 1965, 1966 and 1967.

Niki Lauda
Viele legendäre Rennfahrerkarrieren begannen hinter dem Lenkrad eines Mini. Der 1949 in Wien geborene Andreas Nikolaus Lauda war erst 19 Jahre alt, als er dem damaligen österreichischen Tourenwagenmeister Fritz Baumgartner dessen renntauglichen Mini abkaufte. Im April 1968 ging er in seinem ersten Bergrennen unweit von Linz an den Start und wurde prompt Zweiter. Nur zwei Wochen später bewies er sein Talent mit dem ersten Sieg seiner Laufbahn, die ihn später zu drei Formel-1-Titeln führen sollte. Ebenso wie Niki Lauda sammelten die Formel-1-Champions Graham Hill, Jackie Stewart, John Surtees, Jochen Rindt und James Hunt erste Wettkampferfahrungen im klassischen Mini. Auch Juan Pablo Montoya und David Coulthard fuhren gern darin.

/

Many legendary racing driver careers began at the wheel of a Mini. Andreas Nikolaus Lauda, born in Vienna in 1949, was only 19 years old when he bought former Austrian touring car champion Fritz Baumgartner's racing Mini. In April 1968, he entered his first mountain race not far from Linz and immediately came second. Only two weeks later, he proved his talent by winning the first race of his career which would later lead to three Formula 1 titles. Just like Niki Lauda, other Formula 1 champions such as Graham Hill, Jackie Stewart, John Surtees, Jochen Rindt and James Hunt gained their first racing experience in the classic Mini. Juan Pablo Montoya and David Coulthard also loved racing the Mini.

Stéphane Peterhansel
Er ist der Rekordsieger der Rallye Dakar und wird auch respektvoll »Mister Dakar« genannt: Der 1965 geborene Franzose ist einer der bekanntesten Offroad-Motorsportler und hat bisher bei der härtesten Langstreckenrallye der Welt elf Siege errungen: sechsmal gewann er die begehrte Trophäe mit dem Motorrad, fünfmal in der Autokategorie. In den Jahren 2012 und 2013 holte er den Sieg im MINI ALL 4 Racing.

/

Peterhansel is the record winner of the Dakar Rally and is respectfully known as 'Mister Dakar'. Born in 1965 the Frenchman is one of the best-known off-road racing drivers and has so far won eleven victories in the world's toughest long-distance rallies. He won the coveted trophy six times in the motorcycle category and five times in the car category. In 2012 and 2013, he won the race in a MINI ALL4 Racing.

DESIGN UND TECHNIK
/
DESIGN AND TECHNOLOGY

Exterior Design Die Gestaltung des klassischen Mini ab 1959 erwies sich als modern und kompakt und trug wesentlich zum Erfolg des innovativen Kleinwagens bei. Das Design, das auf Sir Alec Issigonis zurückgeht, war primär ingenieurgetrieben und funktionalen sowie rationalen Kriterien verpflichtet, nach dem Prinzip »form follows function«. Am Anfang stand die Vorgabe, auf kleinster Grundfläche Raum für vier Personen zu schaffen. Daraus ergaben sich bestimmte Proportionen – das wichtigste Kriterium des Designs, – damals wie heute.

Der klassische Mini von 1959 blieb in den folgenden 41 Jahren seiner Grundkonzeption treu und formte über die Jahrzehnte bestimmte Designmerkmale bis hin zu Einzelformen, die mittlerweile Kult geworden sind. Zu einer grundlegenden Überarbeitung oder Weiterentwicklung des Exterior Designs kam es in den Jahren jedoch nicht.

Bei der Konzeption des neuen MINI, der 2000 präsentiert wurde, gingen Designer und Techniker mit besonderer Behutsamkeit vor. Es sollte ein nachvollziehbarer Übergang »vom Original zum Original« geschaffen werden, kein Remake, sondern ein authentischer Nachfolger. Nachdem aber der klassische Mini über die Jahrzehnte unverändert geblieben war, musste der Schritt ins Design des 21. Jahrhunderts entsprechend deutlich ausfallen.

Das Geheimnis des neuen MINI ist, dass er unter denselben funktionalen Gesichtspunkten gestaltet wurde wie der klassische Ur-Mini. Denn es galt auch hier, einem Kleinwagen mit Quermotor auf kleinstmöglicher Grundfläche bei größtmöglichem Innenraum ein zeitgemäßes, authentisches Design zu verleihen.

Zu den zeitlosen Erkennungsmerkmalen eines MINI zählen seine kompakten Proportionen, die Mimik in der Frontgestaltung und ausgewählte Merkmale, die allein der MINI besitzt. Er ist geprägt durch kurze Front- und Hecküberhänge der Karosserie und die Position der Räder an den äußersten Ecken des Fahrzeugs, die seine Gokart-ähnlichen Fahreigenschaften auf den ersten Blick zum Ausdruck bringen. Typisch MINI ist die Dreiteilung der Karosserie in den lackierten Wagenkörper, die durchgehende Fenstergrafik, die als umlaufendes Glasband ausgebildet ist, und das Dach, das darüber frei zu schweben scheint.

Exterior Design The classic Mini first built in 1959 already featured an advanced and compact design, which contributed significantly to the innovative small car's success. Sir Alec Issigonis's design was mainly driven by engineering, and based on functional and rational criteria in line with the principle of 'form follows function'. The initial objective was to create space for four passengers, using the smallest possible footprint. This resulted in certain proportions – the key design criterion – that were applicable then and remain so today.

The classic 1959 Mini remained true to its basic design for the next 41 years, creating over the decades particular design aspects including individual forms that have become iconic. The exterior design, however, did not undergo any essential re-working or further development during this time.

In producing the new-generation MINI in 2000, designers and engineers took great care to respect the design of the original. The intention was to create a clear transition 'from the original to the original', designing an authentic successor rather than a simple re-make. Moving the design to the 21st century however required noticeable changes, considering that the classic Mini's design had not been altered substantially since its launch.

The new MINI's secret lay in the fact that it was designed using the same functional criteria as the classic, original Mini, and it was immune to changing fashions. As before, the requirement was to provide a compact car with a transverse engine on a minimum footprint with the greatest amount of interior space and an authentic, contemporary design.

The MINI's timeless characteristics include its compact proportions, the front section design and selected unique MINI features. It is characterized by short front and rear body bulges and the wheels' position at the outermost corners of the vehicle, highlighting the car's go-kart-style handling at first sight. Typical MINI features are the tripartite paintwork on the car body, the all-round windows designed as an encircling strip of glass, and the roof that seems to be free-floating.

The so-called shoulder line, a straight line running from the headlights above the sidewalls to the rear C-pillar, is another critical aspect

Entstehung eines MINI – Claymodell im Designprozess / The birth of a MINI – clay model used in the design process

Designzeichnung des neuen MINI /
Design drawing for the new MINI

Außerdem ist für das Erscheinungsbild die sogenannte Brüstungslinie entscheidend, eine Gerade, welche von den Frontleuchten oberhalb der Seitenwand bis zur hinteren C-Säule verläuft. Der Abstand zur Dachlinie verringert sich nach hinten, sodass eine Keilform entsteht – ein Gestaltungselement, das die Sportlichkeit des MINI betont. Ebenso unverwechselbar ist seine Front: Rundscheinwerfer und der Kühlergrill in Form eines Hexagons bilden das Gesicht des MINI mit seiner sympathischen Mimik und sind, ebenso wie die Türgriffe und die Schulterlinie, mit edlen Chromelementen besetzt.

Einzelne Elemente des klassischen Mini wurden in der aktuellen Variante neu interpretiert: Der klassische Mini besitzt zwischen Kotflügeln und Karosserie eine diagonal verlaufende Blechfalze, die nach außen gerichtet ist. Ausschlaggebend dafür waren Abläufe in der Produktion. Die außen liegenden Schweißnähte ließen sich einfacher setzen. Der neue MINI greift dieses unverwechselbare Element als eigenständiges Designelement auf und gestaltet es neu: Dort, wo einst die Blechnaht der Seitenwände entlangführte, verläuft heute die Fuge der Motorhaube bzw. der vorderen Seitenbauteile. Neu hinzukamen die Einfassungen der Seitenblinker, die gleichzeitig zur Differenzierung der unterschiedlichen Motorvarianten dienen.

Interior Design Zwar gilt das Prinzip »vom Original zum Original« auch im Innenraum, doch kommen hier andere Rahmenbedingungen zum Tragen: Beim klassischen Mini fiel das Interieur aus Gewichts- und Kostengründen sehr spartanisch aus. Das Fahrzeugkonzept von 1959 zeichnet sich durch innovative Technik, ein erstaunliches Raumangebot und eine faszinierende Straßenlage aus, nicht aber durch seine Innenausstattung. Da Issigonis kein besonderes Augenmerk auf Komfort legte, boten im Lauf der Zeit nicht wenige Spezialfirmen entsprechendes Zubehör an.

Für den neuen MINI aus dem Jahr 2001 wurde mit »Evolution outside – revolution inside!« ein von der Tradition abweichendes Ziel vorgegeben. Damit trug die Marke der technischen Entwicklung, dem Trend zur Individualisierung und dem Verlangen nach zeitgemäßem

Mimik des MINI / The MINI's
facial expressions

of the vehicle's appearance. The roof-line clearance reduces from front to rear, creating a wedge shape that emphasizes the MINI's sporty look. The front section is equally unmistakeable: round headlights and the hexagonal radiator grille lend the MINI a sympathetic face, just like the door handles and the shoulder line that are embellished with elegant chrome elements.

Some of the classic Mini's components were re-interpreted for the current version. The classic Mini has a diagonal sheet-metal beading oriented outwards between the front wheel arches and the front doors. This was a result of the manufacturing process, as the outer welding seams were easier to place. The new MINI adopts this distinctive feature as a discrete design element with a twist: the joint line of the bonnet and the lateral front elements follow the contour of the former sidewall sheet-metal joint. Aprons for the side-repeater indicators serve to differentiate the various engine variants and complement the design.

Interior Design The principle of 'from the original to the original' also applies to the MINI's interior although in this context other generalities come into play. The classic Mini's interior looked very spartan, due to issues of weight and cost. The 1959 car featured innovative technology, an astoundingly spacious interior and tenacious road-holding, but almost no standard equipment. As Issigonis did not particularly focus on comfort numerous aftermarket companies filled that gap, offering accessories as time went on.

However the new-generation MINI introduced in 2001 bucked with tradition in line with its design motto: 'Evolution outside – revolution inside!' In this the brand took account of technical developments, the trend towards personalisation and the demand for advanced comfort features. Today's MINI includes navigation, sound and air-conditioning systems, various driver assistance systems, iPod connectivity and other up-to-date features. The new design also influenced the interior, which became even more spacious and can be fitted with a variety of different materials and equipment options. Such orientation towards individual needs had previously been limited to higher vehicle categories.

Komfort Rechnung. Der heutige MINI integriert Navigations-, Sound- und Klimasysteme, diverse Fahrerassistenzsysteme, eine iPod-Verbindung und andere aktuelle Features. Die Neukonzeption wirkte sich auf den Innenraum aus, der im Zuge dessen noch geräumiger ausfiel und zudem mit einer Vielzahl unterschiedlicher Materialien und Ausstattungsoptionen gestaltet werden kann. Bislang war eine Ausrichtung auf individuelle Bedürfnisse allein Fahrzeugen höherer Klassen vorbehalten.

Ein typisches Element für den MINI ist das runde Zentralinstrument in der Mitte des Armaturenbereichs, unverkennbar inspiriert vom runden Tachometer früherer Modelle. Genau dieses markante Instrument wurde über die Jahre innovativ weiterentwickelt und beherbergt nun Navigations-, Entertainment- und Infotainment-Funktionen. Das Kombiinstrument auf der Lenksäule integriert Tachometer, Tank- und Temperaturanzeige.

Während sich der klassische Mini auf ein einziges Anzeigeninstrument beschränkte, bietet der neue MINI eine Instrumententafel, die an ein Flugzeugcockpit erinnert. Eine kraftvolle, weit gestreckte Linie betont die horizontale Ausprägung der Armaturentafel. Klassische Akzente setzen die runden Luftdüsen und die elliptischen Hauptelemente der Türverkleidung. Zudem unterstützen konvexe Rundungen optisch die harmonische Komfortzone im Interieur.

The round central instrument on the dashboard is a typical MINI feature clearly inspired by the round, centrally mounted speedometer of the classic Mini. This distinctive instrument has been developed significantly over the years, and today accommodates innovative navigation, entertainment and infotainment functions. The combined instrument on the steering column includes the speedometer as well as the fuel and temperature gauges.

While the classic Mini was limited to a single instrument, the new MINI features a far more comprehensive dashboard with some similarities to an aeroplane cockpit. A vigorous, widely stretched line emphasizes the dashboard's horizontal shape. The round air nozzles and the elliptic shape of the main elements of the door-lining define a classic style. In addition, convex round contours endorse the harmonious and comfortable interior design.

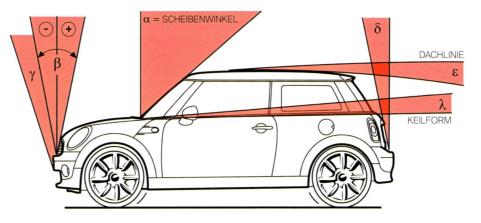

Gestik des MINI / The MINI's gestures

Mit dem klassischen Mini war es dem Konstrukteur Alec Issigonis und seinem Team 1959 gelungen, den Kleinwagen neu zu definieren. Der Mini der ersten Jahrzehnte war ein kleines Auto – 3,05 Meter lang, 1,41 Meter breit und 1,35 Meter hoch –, mit einem Wendekreis von kaum mehr als 10 Metern. Insofern machte Mini seinem Namen alle Ehre. Doch wurde er seit 1959 nicht den Erfordernissen der Zeit angepasst. Er war und blieb ein Klassiker, und seine Erbauer scheuten sich, ihn behutsam weiterzuentwickeln. Der erste Versuch, den Mini zu ersetzen, war der 1980 vorgestellte Austin Mini Metro. Obwohl zwischen 1980 und 1997 mehr als 2 Millionen Modelle verkauft wurden, konnte der Metro dem klassischen Mini nicht den Rang ablaufen.

Ende 2000 wurde der neue MINI erstmals der Öffentlichkeit vorgestellt. Es lag auf der Hand, dass er sich von seinem traditionsreichen Vorgänger und Namensvetter abhob. Schließlich hatten sich die Gegebenheiten seit 1959 fundamental verändert: Der durchschnittliche Erwachsene des 21. Jahrhunderts ist rund 10 Zentimeter größer als sein Vorgänger aus den Fünfzigern. Allein aus ergonomischen Gründen konnten die ursprünglichen Maße nicht beibehalten werden. Entscheidend waren auch die veränderten gesetzlichen Vorschriften und erhöhten Sicherheitsstandards. Crashtests, Emissionen und Recycling forderten ihren Tribut. Der klassische Mini hatte sich niemals einem regulären Crashtest unterziehen müssen.

Im Vergleich zu 1959 sind die Ansprüche der Kunden deutlich gestiegen, die nicht nur mehr Sicherheit, sondern auch mehr Komfort erwarten: Hier seien exemplarisch eine hochwertige Stereoanlage, satellitenbasierte Navigationssysteme, eine Klimaautomatik oder die elektrische Steuerung von Fenstern und Außenspiegeln genannt. Angesichts dieser Anforderungen war es unvermeidlich, dass der neue MINI gegenüber dem klassischen Mini »wachsen« musste. Tatsächlich ist das Modell des 21. Jahrhunderts um 57 Zentimeter länger als sein Vorgänger und bietet 45 Prozent mehr Volumen. Dennoch ist die Verwandtschaft zwischen beiden weiterhin gegeben: Prominente Designmerkmale wurden übernommen, die klassischen Proportionen mit der kurzen Kühlerhaube und dem Rad an jeder Ecke beibehalten.

GROWING UP

In their 1959 design of the classic Mini engineer Alec Issigonis and his team successfully re-defined the concept of the compact car. During the first decades, the Mini was a small car – with a length of 3.05 m, a width of 1.41 m and a height of 1.35 m – with a turning circle hardly exceeding 10 m. In this sense, the Mini lived up to its name. But after 1959 it was no longer suited to the times. The Mini remained a classic, and its maker shied away from any subtle re-development of the car. The first attempt to replace the Mini was the Austin Mini Metro, launched in 1980. Although it sold more than 2 million models between 1980 and 1997, the Metro failed to replace the classic Mini.

The new MINI was presented to the public at the end of 2000. The difference from its classic predecessor and namesake was obvious. After all, the world had changed significantly since 1959: the average 21st-century adult is about 10 cm taller than their 1950s' predecessor. For ergonomic reasons alone the Mini's original dimensions could not been maintained. Changes in statutory requirements and increased safety standards were also critical factors. Crash tests, emissions and recycling requirements took their toll. The classic Mini never had to pass a regular crash test.

Customer demands have increased substantially since 1959 and include higher expectations of both safety and comfort, for example a high-quality stereo system, satellite-based navigation systems, automatic air conditioning or electric windows and door mirrors. In light of these requirements, it was inevitable that the new MINI had to 'grow' by comparison with the classic version. Indeed the 21st-century model is around 57 cm longer than its predecessor and offers 45 per cent more interior volume. Nonetheless, there is still a clear and unmistakeable relationship between the two: important design aspects survived, and the classic proportions with a short bonnet and wheels at the outermost corners remain.

Both car concepts share significant similarities. Both the classic and the new MINI offer maximum interior space on a minimum footprint. While the 1959 Mini managed this in an astonishing way, the 2001 version provided a new interpretation of this complex task:

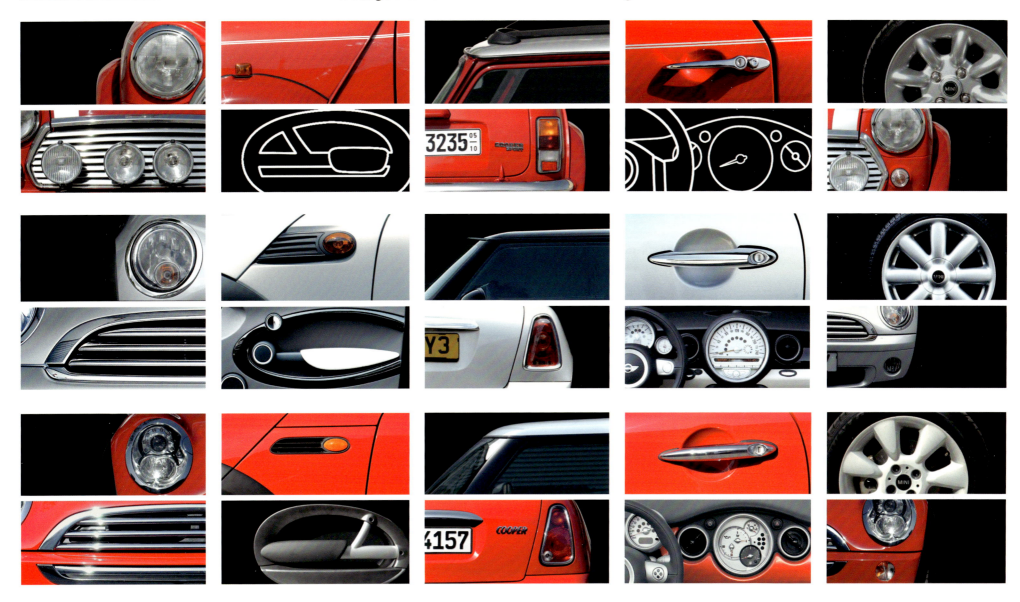

Beide Fahrzeugkonzepte haben wesentliche Gemeinsamkeiten: Der klassische wie der neue MINI bieten maximalen Innenraum auf minimaler Verkehrsfläche. 1959 ist der Plan auf verblüffende Weise gelungen, 2001 folgte die Neuinterpretation dieser kniffligen Aufgabe: Gleich, ob man in einem alten oder neuen MINI sitzt, charakteristisch ist das legendäre »Gokart-Feeling«, das man hinterm Steuer verspürt.

In beiden Fällen ist der MINI innovativ, was den Mut zu neuen Lösungen betrifft, und revolutionär im Umgang mit überkommenen Vorstellungen. Eröffnete der erste MINI die Tür zum Kleinwagensegment, repräsentiert der aktuelle MINI in dieser Automobilkategorie die erste Premiummarke.

whether you drive an old or a new MINI, you still get that characteristic and legendary go-kart feeling.

Both Minis are innovative in terms of daring new solutions and revolutionary in dealing with out-dated concepts. While the first Mini was an entry product to the compact car category, the new MINI is the first premium car in this market sector.

Konzeption Am Beginn des Designprozesses steht der Auftrag, ein neues Fahrzeug zu entwerfen. Ein konzeptioneller Rahmen wird definiert, Anforderungen von Kunden und Markt fließen ein. Diese Anforderungen werden in erste Entwürfe und Charakterskizzen übersetzt. Bis zu zehn Designer aus den Studios in München und von BMW Group DesignworksUSA treten in einen internen Designwettbewerb. Bei regelmäßigen Treffen mit dem Designchef werden Skizzen und Ideen präsentiert und bewertet. Zu den Inspirationsquellen der Designer zählen die Historie von MINI, aber auch Fachmessen und Beispiele aus der Natur, Mode und Architektur sowie aus dem Objekt- und Möbelbereich.

Skizzen Nach der Definition des Charakters und der Festlegung der Ziele beginnt die konkrete Skizzenphase. Jeder Designer im Wettbewerb startet mit einem leeren Blatt Papier. Zahllose Entwürfe entstehen, die Konzepte nehmen langsam Gestalt an. Die ersten Skizzen sind oft nicht mehr als grafische Notizen – die ersten zeichnerischen Spuren einer Idee. Sie sind weder fein ausgearbeitet noch vollkommen, aber genau der unvollständige Charakter macht ihren unvergleichbaren Reiz aus. Über den Prozess hinweg folgen ihnen Zeichnungen, die mit detailreicherer Ausformung die Entstehung des neuen Fahrzeugdesigns klarer erkennen lassen.

Die Kunst der Linie Nach ersten Skizzen erstellen die Designer sogenannte »Tape Drawings« auf einer Art technischer Landkarte des späteren Fahrzeugs mit allen bekannten baulichen Bedingungen. An einer großen weißflächigen Wand wird der Tape Plan, die Darstellung des zukünftigen Fahrzeugs, maßstabsgetreu bzw. in Originalgröße erstellt. Dabei wird schwarzes, flexibles Klebeband, das sogenannte Tape, aufgebracht. Damit lassen sich Proportionen und markante Konturlinien studieren, rasch ändern und festlegen. Beim fertigen Tape Drawing geben bereits Silhouette und Linien eine Vorahnung auf den späteren Charakter des jeweiligen Fahrzeugs.

DESIGNPROZESS / DESIGN PROCESS

Concept Development A design process starts with the commission to develop a new car. The first step is to define a conceptual framework, considering customer and market requirements. These requirements are translated into first drafts and character sketches. Up to ten designers from studios in Munich and the BMW Group DesignworksUSA participate in an internal design competition. Drafts and ideas are presented and assessed during regular review meetings with the Head of Design. The designers seek inspiration from MINI history, but also from trade fairs and examples from nature, fashion and architecture, as well as from the product design and furniture sectors.

Sketches Once the character is defined and the objectives are identified, the designers enter the actual sketching phase. Every designer participating in the competition starts with an empty sheet of paper. With countless drafts being created, the concepts slowly take shape. The first sketches are often simple graphic notes – the first drawn traces of an idea. They are neither highly elaborate nor perfect but this imperfection creates a unique appeal. All through the process these sketches are followed by more detailed and elaborated drawings that illustrate the birth of a new car design.

The Art of the Line Based on the first sketches, the designers prepare so-called 'tape drawings' on a kind of technical road-map for the future vehicle, including all known constructional conditions. The designers create the tape plan on a large white wall, to illustrate the future car true to scale and in life size. A black flexible adhesive tape is used in this process. This enables designers to analyse and define the proportions and distinctive contour lines and change them quickly. The silhouettes and contours of the finished tape drawing provides a first impression of the car's final character.

Form-finding and Creating the Series In addition to the tape drawing, the designer and the form-finding modeller create virtual and real models of the car. CAD (Computer-aided Design) models are especially

Von der Idee zum Modell – der Designprozess bei MINI / From the idea to the model – the MINI design process

Virtuelles Designmodell / Virtual design model

Formfindung und Umsetzung in die Serie Neben dem Tape Drawing entstehen in der Zusammenarbeit von Designer und Formfindungsmodelleur virtuelle und reelle Modelle des Fahrzeugs. CAD-Modelle (Computer Aided Design) eignen sich besonders, um einzelne Bauteile und Komponenten zu betrachten. Will man die reale Gesamtwirkung eines Fahrzeugs beurteilen, bieten sich eher Claymodelle in Originalgröße (1:1) an. Clay ist ein Industriewerkstoff, der sich ähnlich wie Knetmasse verarbeiten lässt. In der Phase der Formfindung geht es an die Ausarbeitung von Flächen, Linien und Details. Die Modelleure formen dabei wahre »Design-Skulpturen«.

Finale Der interne Designwettbewerb dauert nun bereits ein Jahr, und es wurde eine erste Modellvorauswahl getroffen. Im Rahmen einer finalen Präsentation entscheidet der Konzernvorstand über das endgültige Design für Exterieur, Interieur sowie Farbe und Material. Nun folgt die Realisierung: Aus der von Hand geformten Skulptur wird ein von Menschen und Maschinen reproduzierbares Produkt.

suited for viewing individual parts and components. A life-size clay model is preferred for assessing the car's overall appearance. Modelling clay is an industrial material that can be worked like plasticine. The form-finding phase includes the elaboration of surfaces, lines and details, enabling the modellers to create real 'design sculptures'.

The Final Lap A year has passed since the design competition started and an initial pre-selection of models has been made. The group management decides upon the final exterior and interior design, as well as the colour and material during a final presentation. Next, the concept becomes real: the hand-made sculpture is developed into a product reproducible by man and machine.

Entwurfskizze für den / Sketches for the
Superleggera™ Vision, 2014

102

Designzeichnung für den MINI /
Design drawing for the MINI

DESIGN-ZEICHNUNGEN / DESIGN SKETCHES

Der legendäre Ruhm der Marke MINI verdankt sich vor allem dem Design, dem kreativen Vermögen einzelner Persönlichkeiten und dem Zusammenspiel ganzer Teams, ihrem Streben nach optimaler Formgebung als auch ihrem Bemühen um eine behutsame Weiterentwicklung des ersten Mini. Grund genug, an dieser Stelle Designer und Gestalter zu erwähnen, die maßgeblich am Design des MINI teilhatten.

Den Anfang bildet selbstverständlich Sir Alec Issigonis selbst, der nicht nur als Ingenieur und Konstrukteur den Mini in seinem gesamten Charakter geprägt hat. Seine ersten Entwürfe zum Exterieur und Interieur, von denen sich nur weniges erhalten hat, entsprechen modernen Vorstellungen und grundlegenden Designprinzipien des 20. Jahrhunderts. Die Gestaltung des kleinen Kompaktwagens basiert auf einer Funktionalität, die die Bedürfnisse des Kunden im Blick hatte, sowie auf einer kompromisslosen Reduktion auf das Wesentliche. Diese Prinzipien haben das Design des Mini vor allem in den ersten vier Jahrzehnten geprägt. Issigonis zur Seite stand von Beginn an Ricardo »Dick« Burzi. 1968 bzw. 1970 kamen entscheidende Impulse von Roy Haynes und Harris Mann. Seit Mitte der Achtzigerjahre trägt der klassische Mini die Handschrift von Geoff Upex.

Eine bedeutende Rolle in der Designhistorie der Marke spielt Frank Stephenson, der den neuen MINI der Ära ab 2001 maßgeblich prägte. Mit besonderer Hingabe widmete er sich der Gestaltung des großen Ganzen wie auch kleinster Details, betrieb Studien anhand historischer Vorbilder ab 1959 und fand zu einer stimmigen Neuinterpretation des klassischen Mini. Die besondere Herausforderung lag im Brückenschlag zwischen zeitgemäßen Lösungen für Anforderungen wie Komfort und Sicherheit und der Beibehaltung MINI-typischer Designmerkmale und -ikonen. Ihm zur Seite stand Tony Hunter, der die Grundzüge des neuen Interieur-Konzepts verantwortete und dabei mit großem Feingefühl agierte.

Auf den bisher erarbeiteten Studien baute ab 2000 Gert Hildebrand auf, der als Designchef bis 2012 sowohl die zukünftige Designlinie formte als auch die strategische Ausrichtung der Modellpalette vorgab. Er war der kreative Kopf, der eine erfolgreiche MINI-Familie

The legendary fame of the MINI brand owes its success above all to its design, the creative abilities of individual personalities, the co-operation of entire teams, their pursuit of optimal design, and their efforts to create a careful development of the first Mini. All of which provides sufficient reason to mention those designers who have made significant contributions to the design of the MINI.

We begin, of course, with Sir Alec Issigonis, who shaped the Mini and its distinctive character as a whole, and not only as an engineer and manufacturer. His first interior and exterior design sketches, of which very few have survived, echo modern ideas and basic design principles of the 20th century. The creation of the small compact car is based on a functionality that keeps the needs of the customer in mind and uncompromisingly reduces everything to the essentials. These principles characterized Mini design in its first four decades above all. From the outset, Issigonis was assisted by Ricardo 'Dick' Burzi. In 1968 and 1970 respectively, Roy Haynes and Harris Mann added their own crucial input. Since the mid-1980s, the classic Mini bore the signature of Geoff Upex.

A significant role in the brand's design history is played by Frank Stephenson, who has shaped the era of the new MINI considerably since 2001. With singular devotion, he focused on the big picture, seeing design as a whole yet concentrating even on the smallest details, carrying out research based on historical models from 1959 onwards, and coming up with a coherent new interpretation of the classic Mini. This particular challenge lay in creating a bridge between contemporary solutions to modern comfort and safety requirements and maintaining the iconic face of MINI with its signature features. He was actively supported by Tony Hunter, who with great discretion and discernment took responsibility for the basic principles that underpin the new interior.

Building on previous research initiatives, Gert Hildebrand, as Head of Design from 2000 to 2012, helped both to shape the future design direction and also to provide a strategic way forward for the model range. He was the creative head who made a successful MINI

möglich machte. Auch entwickelte er zahlreiche ungewöhnliche Studien sowie Concept-Fahrzeuge und fand ausgefallene Detaillösungen. Ab 2002 bis 2013 lag das Exterior Design der Marke MINI in den Händen von Marcus Syring. In seinen virtuos angelegten Handzeichnungen und Detailskizzen spürt man die besondere Leidenschaft, die es braucht, um der Individualität und dem ästhetischen Anspruch eines MINI gerecht zu werden. Im Zeitraum 2004 bis 2006 trug Marc Girard maßgeblich zur Aufwertung des Interior Design bei.

Seit 2012 verantwortet Anders Warming als Leiter des Designstudios die Geschicke der Marke MINI. Seine Arbeit, die klaren Fokus auf die Historie legt und von teilweise radikalen Ideen geprägt ist, wird die Marke MINI in die Zukunft führen. Der Dreh- und Angelpunkt all seiner Überlegungen ist die Authentizität. So zeigt das Design der Marke heute eine erwachsene und seriöse Formensprache, hochwertige Materialien und formschöne Details ergeben ein stimmiges Gesamtbild. Darüber hinaus nimmt sich MINI wichtigen Themen an, die die Zukunft betreffen. Elektroantriebe werden getestet, ein User Interface Design ist im Einsatz, das den Fahrer und seinen persönlichen Lifestyle unterstützt, und auch die Individualisierungsmöglichkeiten sind unvergleichlich breit gefächert. Im Sommer 2013 übernahm Christopher Weil die Leitung des Exterior Design Teams der Marke MINI und gestaltet seither mit ausgeprägtem Sinn für perfekte Proportionen und formschöne Flächen das Erscheinungsbild der Marke.

Oliver Sieghart verantwortet seit 2006 das Interior Design bei MINI. In dieser Zeit entwickelte er aus einzelnen Designmerkmalen unverkennbare Ikonen des MINI-Design. Eine klare Geometrie und zirkulare Elemente sind bestimmende, MINI-typische Eigenheiten.

Seit 2003 leitet Annette Baumeister das Team des Material- und Farbdesign bei MINI. Ihre Kreativität inspiriert dabei nicht nur das eigene Team, sondern auch Kooperationspartner außerhalb der Automobilbranche. So gab es in den letzten Jahren wiederholt spektakuläre Projekte, etwa mit Ilaria Venturini Fendi oder dem niederländischen Designduo Scholten & Baijings.

family possible. He also developed numerous exceptional studies and concept cars, and came up with unusual, detailed solutions. From 2002 to 2013 exterior design of the MINI brand lay in the hands of Marcus Syring. In his masterful hand drawings and detailed sketches we get a sense of the extraordinary passion needed to satisfy the requirements for individuality and aesthetic criteria demanded by a MINI. Between 2004 and 2006 Marc Girard made a significant contribution to the upgrading of the MINI's interior design.

Since 2012, Anders Warming, Head of the Design Studio, has been responsible for the future of the MINI brand. His work, which places a clear focus on historical aspects and which is characterized by occasionally radical ideas, will lead the MINI brand into the future. In all of his thinking authenticity is of paramount importance. Thus, brand design today displays a mature and distinctive formal language, while high-quality materials and elegant formal details produce a coherent overall image. Beyond this, MINI is also addressing important issues for the future. Electric motors are being tested, a User Interface Design is in use which supports drivers and their own personal lifestyle, and there is also an unparalleled range of individualisation options available. In summer 2013 Christopher Weil became Head of the Exterior Design teams for the MINI brand and has since made a significant impact on the brand's image with his particularly well-developed eye for perfect proportions and beautifully designed surfaces.

Oliver Sieghart has been responsible for Interior Design at MINI since 2006. During this time he has developed the unmistakable icons of MINI design out of individual design elements. A clear geometry and circular features are the hallmarks typical of the MINI.

Since 2003 Annette Baumeister has led the Colour and Trim Design team at MINI. Her creativity is not only an inspiration to her own team but also to co-operation partners outside the automobile industry. In recent years this had led to a number of spectacular projects, for example in collaboration with Ilaria Venturini Fendi or the Dutch design duo Scholten & Baijings.

MINI ACV 30

Adrian van Hooydonk, 1995

MINI Hatch

Frank Stephenson, 1997

MINI Hatch

Tony Hunter, 1997

MINI Clubman

Bruno Amatino, 2003

MINI Cabrio Cooper

Marcus Syring, 2004

DACHKONZEPT

MINI Concept Geneva

Dirk Müller-Stolz, 2005

MINI Concept Geneva

Dirk Müller-Stolz, 2005

MINI Clubman

Henning Holstein, 2006

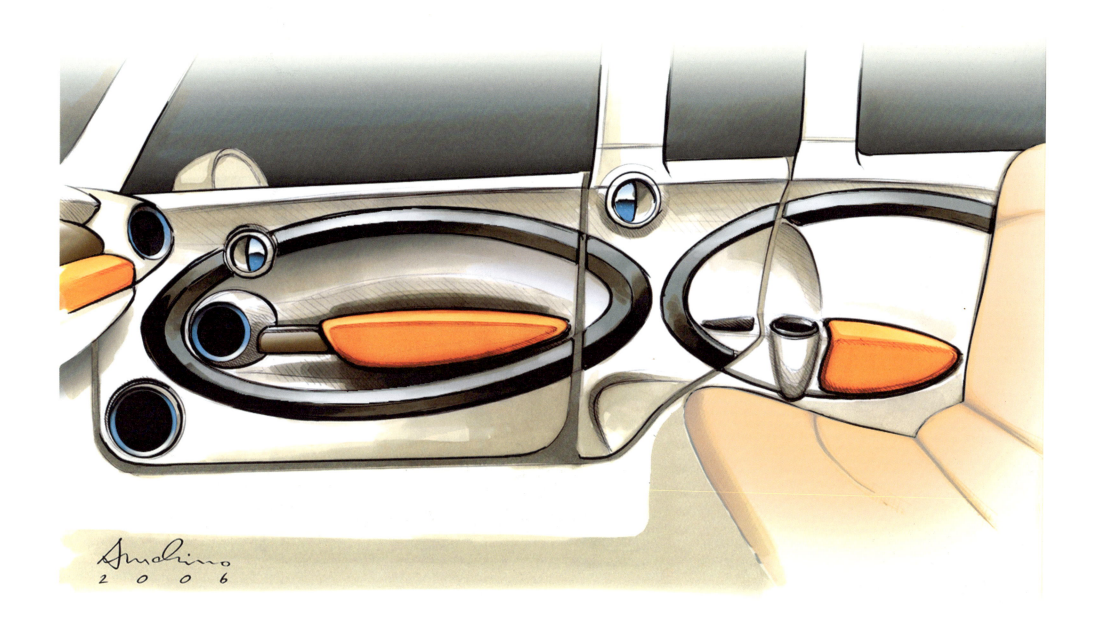

MINI Clubman

Bruno Amatino, 2006

MINI Clubman

Fabien Clottu, 2007

MINI Clubman

Marcus Syring, 2007

MINI Clubman

Marcus Syring, 2007

MINI Beachcomber

Fabien Clottu, 2009

MINI Rocketman Concept

John Buckingham, 2010

MINI Rocketman Concept

John Buckingham, 2010

MINI Coupé

Marcus Syring, 2011

MINI Concept Coupé

Henning Holstein, 2011

MINI Superleggera™ Vison Concept

Louis de Fabribeckers, 2013

MINI Superleggera™ Vison Concept

Thomas Wu, 2014

MINI Clubman Concept

Alexander Schneider, 2014

Die Anfänge der Automobilproduktion in Oxford reichen rund hundert Jahre zurück: 1913 hatte William Morris im Vorort Cowley das Werk Morris Motors gegründet. Seitdem sind auf diesem Areal rund 11 700 000 Fahrzeuge vom Band gelaufen, und siebenmal haben die Fabrikhallen den Besitzer gewechselt: Mit dem Anschluss von Morris an Austin 1952 entstand die British Motor Corporation, die nach der Vereinigung mit Jaguar 1966 wiederum zur British Motor Holdings wurde. Schließlich kamen noch Leyland Motors hinzu, und es entstand die Leyland Motor Corporation. Ab 1986 hieß der Besitzer Rover Group, 1988 British Aerospace und nunmehr – seit 1994 – sind die Liegenschaften im Besitz der BMW Group.

Höhen und Tiefen haben die hier Beschäftigten miterlebt: Mit dem Wechsel in der Firmenleitung verband sich die Hoffnung auf dringend benötigte Investitionen und die Entwicklung absatzstarker Fahrzeugkonzepte. Neue Modelle wurden aufgelegt, einige jedoch schon nach kurzer Zeit wieder aus dem Verkehr gezogen. Voller Stolz erinnern sich ehemalige Werksangehörige an die Produktion des in Großbritannien bis heute geliebten Morris Minor oder an die Sonderfertigung des Hindustan Ambassador ab 1958, den Morris Oxford MO, eine viertürige Limousine, die in Indien bis 2014 in Lizenz gebaut wurde und dort heute noch das Straßenbild bestimmt. Unvergessen ist auch die Episode der Honda-Fabrikation in den Achtzigerjahren, der Bau des Morris Marina oder des Austin Maestro.

In den Sechzigerjahren waren im Werk Oxford etwa 28 000 Arbeiter beschäftigt, heute sind es rund 4000. Wo ab 1913 etwa zwanzig Autos pro Woche fertiggestellt wurden, verlassen heute an einem einzigen Arbeitstag bis zu 1000 Premiumfahrzeuge der Marke MINI die Werkshallen. Ein besonderer Tag war zweifellos der 8. Mai 1959, als der erste Morris Mini-Minor gefertigt wurde. Einen Monat zuvor, am 4. April 1959, war der fast baugleiche Austin Seven im Automobilwerk in Longbridge, dem Industriestandort bei Birmingham, vom Band gegangen. Die Zwillinge wurden der Öffentlichkeit am 26. August 1959 präsentiert. Die Produktion des ersten Mini, wie er erst später heißen sollte, verteilte sich also auf zwei getrennte Standorte, wobei es bei Kapazitäts-

PRODUKTION / PRODUCTION

Automotive production at the Cowley factory in Oxford started some 100 years ago when William Morris established the Morris Motors production plant in 1913. Since then, this plant has produced around 11,700,000 cars under seven owners. Morris joined Austin to become the British Motor Corporation in 1952, which merged with Jaguar to become the British Motor Holdings in 1966, and was finally merged with Leyland Motors (Rover-Triumph) to become the British Leyland Motor Corporation. BL was renamed the Rover Group in 1986, which was bought by British Aerospace in 1988 and finally, in 1994, by the BMW Group.

The workforce has experienced highs and lows. The new management raised expectations of urgently needed investment and the development of profit-making cars. New models were released, although some were quite quickly withdrawn from the market. Former employees proudly remember the production of the Morris Minor, which is still popular today in Great Britain; or the Morris Oxford MO, which would live on as the Hindustan Ambassador series produced from 1958 – a four-door saloon that was built in India until 2014 and still dominates Indian roads. The Honda production during the 1980s is also fondly remembered, as is work on the Morris Marina and the Austin Maestro.

In the 1960s the Oxford plant had a headcount of 28,000, while today it employs some 4,000 workers. When production started in 1913, around 20 cars left the plant each week. Now, the plant produces up to 1,000 premium MINIs on a single working day. A particularly memorable occasion was 8 May 1959, the day the first Morris Mini-Minor was completed. The almost identical Austin Seven had rolled off the assembly line at the production plant in Longbridge, an industrial site near Birmingham, only a month before on 4 April 1959. These twins were introduced to the public on 26 August 1959. The production of the first Mini-to-be started out on two separate sites, although capacity issues resulted in a certain amount of 'intermixing'. While Oxford produced the standard model of the classic Mini, all other body variants, such as the Mini Van, the Mini Pick-up or even the Morris Mini Traveller and the Austin Seven Countryman were built in Longbridge. One

Produktion / Production in Longbridge

problemen sehr wohl zu »Mischungen« kam. Während in Oxford das Standardmodell des klassischen Mini vom Band ging, stammten alle weiteren Karosserievarianten – der Mini Van, der Mini Pick-up bis zum Morris Mini Traveller und Austin Seven Countryman – aus Longbridge. Hauptsächlich in Oxford wurde ein Drittel aller produzierten Fahrzeuge in Form zerlegter Bausätze vorbereitet, die in die fernen Werke des BMC-Imperiums exportiert wurden. An Ort und Stelle, beispielsweise in Australien oder Südafrika, erfolgte dann die Endmontage.

Erst zehn Jahre später, im Jahr 1969, wurde die gesamte Mini-Produktion nach Longbridge verlegt, da man in Oxford Platz für den Austin Maxi schaffen musste. Die damalige Logistik stellte eine besondere Herausforderung dar, waren beide Fabriken doch etwa 100 Kilometer voneinander entfernt und eine moderne Autobahn noch nicht gebaut. Ein engmaschiges Netz an Zulieferern versorgte die Produktionsstätten mit Rohkarosserien, Bauteilen und Motoren.

Die sorgfältig konstruierte Rohkarosserie des Mini wog damals nur 140 Kilogramm und besaß eine vorbildliche Torsionssteifigkeit. Dafür sorgten in Längsrichtung zwei Schweller, ein leichter Tunnel in der Wagenmitte, der die Abgasanlage aufnahm, sowie die Radkästen. In Querrichtung stabilisierten die robuste Spritzwand zwischen Motorraum und Fahrgastzelle, eine Quertraverse unter den Vordersitzen und die Kofferraumwand.

Die Produktionsstatistiken zeichnen ein klares Bild vom wechselhaften Erfolg des MINI über die Jahre: Zählte man im ersten vollen Produktionsjahr 1960 insgesamt 116 677 Mini-Fahrzeuge, waren es 1967 – dem erfolgreichsten Jahr überhaupt – 326 818. Doch schon 1978 sank die Zahl auf unter 200 000, 1981 auf unter 100 000, 1993 gar auf unter 25 000. Im Jahr nach der großen Modellpflege von 1996 erwies sich das Produktionsvolumen von knapp 15 000 Exemplaren als alarmierend. Die Tage des klassischen Mini waren gezählt. Nicht wenige Experten hatten schon 1986, als der fünfmillionste Mini die Werkshallen verließ, zu einem baldigen Produktionsstopp geraten. Doch wer wollte das Kultauto mit dem liebenswerten Image zu Grabe tragen? 1999 fiel die endgültige Entscheidung für das Aus in Longbridge. Nach 41 Jahren

third of all manufactured cars were processed as CKD (Completely Knocked Down) kits mainly in Oxford and shipped to the international BMC sites. The final assembly was performed on site in Australia, South Africa, Malta or New Zealand, for example.

The entire UK Mini production was only re-located to Longbridge ten years later in 1969, to provide capacity for the Austin Maxi in Oxford. Logistics were a particular challenge in those days, as the two factories were some 100 km apart, and modern motorways had not yet been built. A close-knit network of suppliers provided the production plants with raw bodies, components and engines.

Back then the carefully designed Mini bodyshells weighed only 140 kg and featured excellent torsional rigidity, which was created by two longitudinal sillboards and a slight tunnel in the centre of the body accommodating the exhaust system and the wheel cases. Lateral stability was maintained by the robust bulkhead between the engine and the passenger compartment, the cross-beam beneath the front seats and the rear boot wall.

Production statistics clearly illustrate the MINI's varied fortunes over time. While the first full production year – 1960 – recorded 116,677 Minis, output reached 326,818 in 1967, the most successful year ever. However, production figures declined in 1978 to under 200,000, fewer than 100,000 in 1981, and even below 25,000 in 1993. The year after the extensive model upgrades in 1996 recorded an alarming all-time low of just 15,000 cars. The days of the classic Mini were numbered. In 1986 numerous industry experts had already recommended a short-term halt in production when the 5 millionth Mini rolled off the assembly line. But who wanted to put an end to an iconic and well-loved car? The final decision to close down production at Longbridge was made in 1999. After 41 years the last classic Mini – number 5,378,776 – left the plant on 4 October 2000. It was given on permanent loan to the British Motor Industry Heritage Trust in Gaydon.

The intention had been to build the new MINI in Longbridge, but when BMW sold the factory to the MG Rover Group in 2000, it retained MINI and re-located the new MINI production to Oxford in a programme

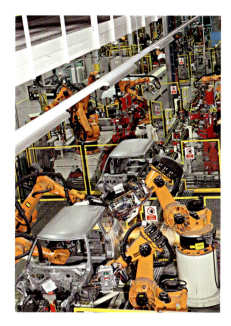

Produktion / Production
in Oxford

verließ dort am 4. Oktober 2000 der letzte klassische Mini – es war der 5378 776ste – das Werk. Als Dauerleihgabe wurde er dem British Motor Industry Heritage Trust in Gaydon vermacht.

Zunächst hatte man die Absicht gehabt, den neuen MINI in Longbridge zu bauen. Als BMW das Werk jedoch 2000 an die MG Rover Group verkaufte, verblieb dort nur der alte Mini, während die Produktion des neuen MINI innerhalb von weniger als 13 Monaten nach Oxford verlegt wurde. Die Übersiedlung der kompletten Produktionsanlage inklusive Robotern, Fließbändern und Brückenteilen stellte für die damals rund 2500 Mitarbeiter in Oxford und ihre Kollegen aus dem BMW-Werk in Regensburg, die den Modernisierungsprozess begleiteten, eine besondere Herausforderung dar. Aber der Erfolg war alle Mühe wert – der Launch-Termin konnte eingehalten werden: Am 26. April 2001 lief der erste neue MINI in der Endmontage vom Band.

Heute verfügt das MINI-Werk in Oxford über eine der modernsten Automobilfertigungsstätten der Welt, ausgestattet mit hochwertiger Fabrikationstechnik: Etwa 1000 neue Roboter wurden in einer der modernsten Karosseriebauten und bestehenden Anlagen installiert. Mithilfe eines Laser-Mess-Systems kann jede Karosserie auf 0,05 Millimeter genau geprüft werden. Auch die Lackiererei, die einen präzisen und umweltschonenden Auftrag der Farbschichten garantiert, entspricht modernsten Qualitätsstandards. Im Herbst 2006 wurde das sogenannte Produktionsdreieck verwirklicht: Neben den Fertigungsstätten für Karosseriebau, Lackierung und Montage in Oxford werden im 70 Kilometer entfernten Swindon Blechteile gefertigt und Karosseriebauteile vormontiert. Unweit von Birmingham befindet sich das Motorenwerk von Hams Hall, dort werden heute Drei- und Vierzylindermotoren gefertigt.

Die Produktion in diesen Werken ist nachweislich auf Erfolgskurs: In Oxford arbeiten heute rund 4000 Mitarbeiter. Voller Stolz bemerkt der Werksleiter Frank Bachmann, sein Werk sei eine der effizientesten, modernsten und flexibelsten Produktionsstätten der Welt: »Wir alle wollen dem MINI das mitgeben, was ihn auszeichnet: eine besondere Qualität in der Verarbeitung und ein unendliches Individualisierungsangebot für unsere Kunden.«

that took no more than 13 months. Re-locating the entire production facilities including robots, conveyor belts and bridge parts proved to be a particular challenge for the 2,500 or so employees in Oxford at this time and their colleagues from the BMW plant in Regensburg, who supported the modernisation process. However, it was all worth the effort, as they met the launch deadline and the first new MINI left the final assembly line on 26 April 2001.

Today the Oxford MINI factory is one of the world's most advanced production plants, equipped with superior manufacturing technology. Some 1,000 new robots have been installed on one of the most advanced body manufacturing lines and existing plants. A laser measuring system enables each car body to be inspected with an accuracy of 0.05 mm. The paint shop, which guarantees precise and sustainable paintwork, also meets the most up-to-date quality standards. The so-called production triangle was introduced in autumn 2006. The Swindon plant produces sheet-metal parts and pre-assembles car body parts alongside the car body, paint shop and assembly activities which are based in the Oxford production plant. Today the Hams Hall engine factory near Birmingham supplies three- and four-cylinder engines.

Production in these plants is clearly on the road to success. Today some 4,000 employees work in Oxford. Managing Director Frank Bachmann proudly asserts that his factory is one of the most efficient, modern and flexible production plants in the world: 'We all want to bring to MINI those features that truly set this car apart: a special quality of workmanship and infinite individualization possibilities for our customers'.

Produktion / Production in Longbridge

Produktion / Production in Oxford

Der Mini seit 1959 Als Leonard Lord, der Chef des Konzerns British Motor Corporation, Ende 1956 den offiziellen Auftrag zum Bau eines »richtigen Kleinwagens« gab, war eine seiner Bedingungen, einen Motor aus der laufenden Produktion zu verwenden. Denn BMC konnte zu diesem Zeitpunkt – wie andere Automobilhersteller auch – nur auf sehr begrenzte Finanzmittel zurückgreifen.

Als Triebwerk kam somit nur der sogenannte Serie-A-Motor in Frage, ein leistungsstarker Vierzylindermotor, der bereits den Austin A30 und den legendären Morris Minor angetrieben hatte, ein Motor, der schon in über einer halben Million Einheiten produziert und ständig verbessert worden war. Alec Issigonis und sein Team überarbeiteten diesen Motor und stellten nach einigen Tests fest, dass die gesamte Motoreinheit um 180 Grad gedreht werden musste, sollte sie in dem beengten Motorraum Platz finden. Zunächst brachte es das neue Aggregat bei einem Hubraum von 948 ccm auf eine Leistung von 37 PS. Doch dieser Wert war für Fahrwerk und Bremsen zu viel, sodass der Hubraum auf 848 ccm und die Leistung auf 34 PS bei 5500 Umdrehungen pro Minute verringert wurde – eine Nenndrehzahl, die damals außergewöhnlich hoch war und nur von hochkarätigen Sportmotoren erzielt wurde.

Zu den Besonderheiten gehört zweifellos der quer montierte Einbau des Vierzylinders, ebenso aber die erstmalige Platzierung des Getriebes unter dem Motor, direkt zwischen den Vorderrädern. Dazu musste die Maschine nur einige Zentimeter höher positioniert werden. Motor und Getriebe bekamen einen gemeinsamen Ölkreislauf, womit Raum gewonnen wurde für den seitlich angebrachten Kühler, ebenso für Lenkung und Nebenaggregate.

Der Mini seit 2000 Die erste Motorengeneration des neuen MINI, der 2000 vorgestellt wurde, wird von einem 1,6-Liter-»Pentagon«-Grundmotor angetrieben, der gemeinsam von BMW und Chrysler entwickelt und in Brasilien gebaut wurde.

Der MINI One wird von einer Motorvariante mit 66 kW/90 PS angetrieben, beim MINI Cooper beträgt die Leistung 85 kW/115 PS. Damit

DER MOTOR
/
THE ENGINE

The Mini after 1959 When Leonard Lord, Head of the British Motor Corporation Group, officially commissioned the construction of a 'real small car', one of his stipulations was the use of an existing engine already in production. Like other automotive manufacturers at this time BMC had only limited financial production capacity its disposal.

The so-called A-Series engine – a four-cylinder engine already used in the Austin A30 and the legendary Morris Minor, of which more than 500,000 units had been produced and which had been subject to continuous improvements – was thus the preferred option. Alec Issigonis and his team revised this engine, realizing after a few tests that the engine block had to be mounted transversally for optimum packaging; further testing then demanded it be rotated by 180 degrees for the best reliability. The first modified new engine provided a cubic capacity of 948 cc and an output of 37 bhp. However this was too powerful for the chassis and braking system, so it was decided to reduce the cubic capacity to 848 cc and the output to 34 bhp at 5,500 rpm – an extraordinarily high engine speed back then and one only delivered by high-profile sports engines.

Distinctive features undoubtedly included the transverse-mounted four-cylinder engine, and the gearbox located in the sump under the engine – an industry first. This was achieved by lifting the engine just a few centimetres. Engine and gearbox shared their oil and room was made for the side-mounted radiator, the steering system and ancillaries.

The MINI after 2000 The first new MINI, presented in 2000, featured a 1.6-litre 'Pentagon' basic engine developed jointly by BMW and Chrysler, and manufactured in Brazil.

While the MINI One is equipped with a 66 kW/90 bhp engine, the MINI Cooper engine delivers 85 kW/115 bhp. The new MINI is able to run at 185 km/h and to accelerate from 0 to 100 km/h in 10.9 seconds. Its consumption is no more than 6.5 litres per 100 km. The engine meets the EU4 emission requirements and is among the few available engines not requiring a secondary air injection or exhaust gas recirculation system.

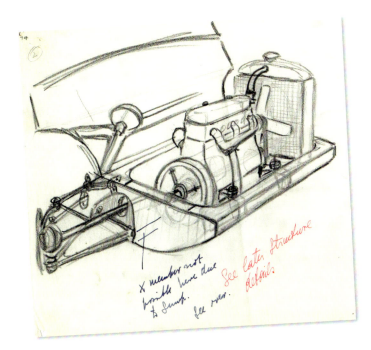

Motorskizze von / Sketch for an engine by
Sir Alec Issigonis

sprintet der neue MINI in 10,9 Sekunden von 0 auf 100 km/h und wird
185 km/h schnell. Der Verbrauch liegt bei geringen 6,5 Litern auf
100 Kilometern. Der Motor erfüllt die EU4-Abgasbestimmungen und ist
einer der wenigen Motoren auf dem Markt, der keine Sekundärluftein-
blasung oder Abgasrückführung benötigt.

Weitere Motorvarianten wurden ins Programm aufgenommen, so
der MINI Cooper S, der im Juni 2002 mit einem 120 kW/163 PS starken
Kompressormotor als exklusive Fahrmaschine an den Start ging, oder
der MINI One D, der 2003 als erstes Dieselmodell in der Geschichte der
Marke neue Maßstäbe für Wirtschaftlichkeit setzte.

Der Mini seit 2006 Die zweite Generation des MINI seit der Neukonzeption von 2001, die 2006 vorgestellt wurde, zeichnete sich unter anderem durch neue, kraftvollere und zugleich erheblich effizientere
Motoren aus. Diese ermöglichten es, gemeinsam mit der weiterhin op-
timierten Fahrwerkstechnik den für MINI typischen Fahrspaß zu steigern. Die zur Markteinführung verfügbaren Modelle MINI Cooper S
mit 128 kW/175 PS und MINI Cooper mit 88 kW/120 PS begeisterten
durch ihre Fahrleistungen bei gleichzeitig deutlich reduzierten Ver-
brauchs- und Emissionswerten. Die Motoren wurden zusammen mit
PSA (Peugeot/Citroën) entwickelt. Der 1,6-Liter-Vierzylindermotor des
MINI Cooper S ist mit einem Twin-Scroll-Turbolader sowie Benzin-
Direkteinspritzung ausgestattet und erhielt mittlerweile achtmal in
Folge die Auszeichnung des Engine of the Year Award in seiner Klasse.
Im Abgaskrümmer und im Turbolader sind die Kanäle von jeweils zwei
Zylindern voneinander getrennt. Diese Anordnung begünstigt das

Additional engine variants complemented the programme, such as the
MINI Cooper S, which was introduced in June 2002 with an exclusive
120 kW/163 bhp compressor engine, or the MINI One D, which set new
efficiency standards as the brand's first diesel model in 2003.

The MINI after 2006 The second generation of MINI since the re-launch
in 2001 was introduced in 2006 and offered a new, more powerful and
significantly more efficient engine among other distinctive features.
Combined with the improved chassis set-up and body engineering,
these features enabled even more driving pleasure typical of the MINI.
The launch of the MINI Cooper S with 128 kW/175 bhp and the
MINI Cooper with 88 kW/120 bhp offered an amazing driving performance with significantly reduced fuel consumption and emissions. The
engines were developed in co-operation with PSA (Peugeot/Citroën).
The MINI Cooper S's 1.6 litre four-cylinder engine is equipped with a
twin-scroll turbo charger and direct fuel injection, which received the
Engine of the Year Award in its category eight times in a row. The channels of two cylinders respectively are separated in the exhaust manifold and the turbo charger. This layout supports the turbo charger's
response and enables a highly spontaneous power build-up.

With the launch of the MINI One (70 kW/95 bhp) and MINI Cooper D
(80 kW/109 bhp) in 2007, the MINI One (55 kW/75 bhp) and MINI One D
(66 kW/90 bhp) in 2009, and the MINI Cooper SD (105 kW/143 bhp) in
2011, this MINI generation included a total of four petrol and three
diesel models over the last three years. In addition, the stand-alone
MINI John Cooper Works (155 kW/211 bhp) was introduced in 2008
for the first time. The optimised air in- and out-feed and the adjusted
twin-scroll turbo charger in particular underpinned this increase in
performance. The turbo charger features a turbine made of highly
superior material. Furthermore, the maximum boost pressure was
increased from 0.9 to 1.3 bar overpressure. The maximum torque of
260 N m is possible even at 1,850 rpm. For short periods torque rises
to 280 N m when accelerating, as it overboosts between 1,950 and
5,500 rpm.

Motor des Morris Mini-Minor / Morris Mini-Minor engine

Motor des John Cooper Works /
John Cooper Works engine

Ansprechverhalten des Turboladers und ermöglicht einen besonders spontanen Leistungsaufbau.

Mit der Einführung der Modelle MINI One (70 kW/95 PS) und MINI Cooper D (80 kW/109 PS) im Jahr 2007, MINI One (55 kW/75 PS) und MINI One D (66 kW/90 PS) im Jahr 2009 sowie MINI Cooper SD (105 kW/143 PS) im Jahr 2011, war der MINI dieser Generation in den letzten drei Jahren in insgesamt vier Benzin- und drei Dieselvarianten verfügbar. Darüber hinaus wurde ab 2008 erstmals auch der eigenständige MINI John Cooper Works mit einer Leistung von 155 kW/211 PS angeboten. Die Leistungssteigerung erfolgte insbesondere durch eine optimierte Zu- und Abfuhr von Luft sowie durch eine Anpassung des Twin-Scroll-Turboladers. Er verfügt über eine Turbine aus besonders hochwertigem Material. Zudem wurde der maximale Ladedruck von 0,9 auf 1,3 bar Überdruck angehoben. Bereits bei 1850 min^{-1} steht das maximale Drehmoment von 260 Newtonmetern zur Verfügung. Beim Beschleunigen wird das Drehmoment durch eine kurzzeitige Erhöhung des Ladedrucks im Drehzahlbereich zwischen 1950 und 5500 min^{-1} sogar auf 280 Newtonmeter angehoben.

Der Mini seit 2014 Die dritte Generation des MINI, die 2014 vorgestellt wurde, war nicht nur mit einer deutlichen Überarbeitung des Designs, sondern auch mit neuem Antrieb verbunden. Erstmals bot die Marke Drei- und Vierzylindermotoren an, die sich durch die MINI-TwinPower-Turbo-Technologie auszeichnen. Bei allen Ottomotoren umfasst sie eine Turboaufladung, eine Benzin-Direkteinspritzung sowie die variable Nockenwellensteuerung auf der Einlass- und Auslassseite. Bei den Antriebseinheiten der Modelle MINI Cooper S und MINI Cooper kommt eine vollvariable Ventilsteuerung in Form der VALVETRONIC hinzu. Die Dieselmotoren des neuen MINI One D und des neuen MINI Cooper D verfügen über ein Aufladesystem mit variabler Turbinengeometrie sowie über eine Common-Rail-Direkteinspritzung. Alle diese Modellvarianten stehen für verbesserte Motor- und Fahrleistungen. Gegenüber den Vorgängermodellen konnten ihre Verbrauchs- und Emissionswerte um bis zu 27 Prozent reduziert werden.

The MINI after 2014 Introduced in 2014, the third MINI generation was characterized by both remarkable design changes and a new engine. For the first time MINI offered three- and four-cylinder engines with MINI-TwinPower turbo technology. In all petrol engines the technology includes turbo charging, direct fuel injection and a variable camshaft control on the inlet and outlet side. The MINI Cooper S and MINI Cooper engines also benefit from a fully variable VALVETRONIC valve control. The new MINI One D and MINI Cooper D diesel engines feature a charging system equipped with variable turbine geometry and a common rail direct injection. All these models offer improved engine performance and handling. By comparison with their predecessors their fuel consumption and emissions have been successfully reduced by up to 27 per cent.

133

DER NEUE MINI
/
THE NEW MINI

1994 Auf Betreiben des damaligen Vorstandsvorsitzenden Bernd Pischetsrieder erwarb die BMW AG im Januar die Mehrheit der Rover Group mit den Marken MG, MINI und Land Rover. Kurze Zeit später wurde der noch mitbeteiligte Honda-Konzern durch die Übernahme des vollständigen Aktienpakets ausgelöst. Die von Rover bereits beschlossene Produktionseinstellung des bisherigen Mini wurde von BMW durch weitere Investitionen und Hilfsmittel in die britische Firma verhindert. Einerseits, so die Strategie, sollte der neue MINI in Rovers Entwicklungshochburg Gaydon entwickelt und in Longbridge produziert werden, andererseits sollte der klassische Mini bis dahin eine letzte umfassende Modellpflege erhalten.

1995 Die Entscheidung fiel, eine komplette Neuversion des MINI zu entwickeln.

1996 Unter BMW-Ägide wurde der klassische Mini gründlich überarbeitet. Die Türen bekamen einen Seitenaufprallschutz, das Lenkrad erhielt einen Fahrerairbag, hinzu kamen Gurtstraffer und eine Teleskoplenksäule. Das Innere wartete mit einem neuen Armaturenbrett auf. Kunden hatten jetzt mehr Möglichkeiten, ihren MINI nach ihren Bedürfnissen gestalten zu lassen.

1997 Bei der Rallye Monte Carlo und auf dem Genfer Automobilsalon wurde der ACV 30 als neues, sportliches Mini Concept vorgestellt. Die Rover Group ihrerseits präsentierte ihr neues Concept Car in Form des Spiritual.

Auf der IAA in Frankfurt zeigte BMW den ersten Prototypen des neuen MINI und kündigte den Serienanlauf für das Jahr 2000 an.

1998 In den Neunzigerjahren hatte Rover regelmäßig Sondermodelle herausgebracht. 1998 erschien der Mini Paul Smith und setzte die Tradition attraktiver Sondermodelle fort. Insgesamt sollten in der klassischen Ära des Mini noch einmal sieben Sonderausführungen entstehen.

1999 Im Modelljahr 1999/2000 sank die Mini-Produktion auf ein Rekordtief von nur noch 7069 Fahrzeugen.

2000 Im März trennte sich BMW sechs Jahre nach der Übernahme von den Marken Rover, MG und Land Rover: Kurz vor der Hauptversamm-

DER WEG ZUM NEUEN MINI
/
THE WAY TO THE NEW MINI

1994 In January at the instigation of the then CEO Bernd Pischetsrieder, BMW AG acquired a majority shareholding in the Rover Group, along with the MG, Mini and Land Rover brands. A short time later, the company's shareholding in the Honda group was dissolved with the takeover of the complete share package. Rover had already decided to cease production of the classic Mini, but this was halted when BMW invested further resources in the British company. The strategy on the one hand was to create a completely new MINI at Rover's development centre in Gaydon and manufacture it in Longbridge, and on the other to keep the classic Mini and give it a final, comprehensive model makeover.

1995 The decision was made to develop a completely new version of the MINI.

1996 Under the auspices of BMW, the classic Mini underwent a complete re-vamp. The doors were reinforced with side impact protection, the steering wheel was given a driver airbag, and belt tensioners and a telescopic collapsible steering column were also added. The interior boasted a revised dashboard. Customers now had more options when it came to designing their MINI to match their needs.

1997 A new sporting Mini concept, the ACV 30, was unveiled at the Monte Carlo Rally and at the Geneva motor show; Rover Group unveiled a new concept car to the public in the shape of the Spiritual.

At the IAA in Frankfurt, BMW presented the first prototypes of the new MINI and announced the start of series production for 2000.

1998 During the 1990s, Rover had regularly brought out special models. In 1998, the Mini Paul Smith appeared, continuing the tradition of attractive special models. In total, a further seven special models were produced during the classic Mini era.

1999 During the 1999/2000 model year, Mini production dwindled to a record low of just 7,069 cars.

2000 In March, and after just six years together, BMW severed its links with Rover, MG and Land Rover brands. Just prior to the AGM, the Supervisory Board Chairman Joachim Milberg announced that Land Rover was being sold to Ford, and MG Rover to the Phoenix

lung verkündete der Aufsichtsratsvorsitzende Joachim Milberg, dass Land Rover an Ford und MG Rover an die Investorengruppe Phoenix verkauft werden sollte. Allein die Marke MINI verblieb in der BMW Group.

Im September wurde der neue MINI auf dem Pariser Autosalon vorgestellt. Das letzte Modell des klassischen Mini rollte am 4. Oktober 2000 im Werk in Longbridge vom Band. Seine Produktionsnummer lautete 5 387 862. Damit ging eine 41-jährige Ära zu Ende.

2001 Im April lief die Serienproduktion im neuen Werk in Oxford an.

Die britische Presse hatte die Übernahme der Rover Group durch BMW zunächst kritisch gesehen. Immerhin bildeten die Marken Rover, MG, Land Rover und MINI eine letzte Bastion des einst weltweit führenden britischen Automobilbaus. Die Neubelebung des Mini ist aus heutiger Sicht gelungen und darf als eine große Erfolgsgeschichte britisch-bayerischer Zusammenarbeit gelten. Die hohe Bedeutung von MINI in Verbindung mit dem Produktionsstandort in Oxford brachte Großbritanniens Premierminister David Cameron im Juni 2011 auf den Punkt: »Das MINI-Werk in Oxford ist eine unserer großartigen Fertigungserfolgsgeschichten. Das Unternehmen kann auf die erzielten Erfolge stolz sein. Es hat wieder einmal bewiesen, dass Großbritannien eine wichtige Rolle in der weltweiten Automobilindustrie spielt.« Und fügte zwei Monate später hinzu: »Es ist ein Privileg, diese britische Ikone heute von der Fertigungsstraße fahren zu dürfen. Der hier in Oxford gefertigte zweimillionste MINI ist ein Sinnbild für die starke Position Großbritanniens in der weltweiten Automobilindustrie und für die Erfolgsgeschichte der britischen Fertigungsindustrie. Die fortgesetzten Investitionen von BMW zusammen mit den Exporterfolgen – das heißt der Vertrieb in mehr als 90 Länder weltweit – trägt wesentlich zu der Neubelebung der britischen Automobilindustrie bei und ist ein hervorragendes Beispiel für das angestrebte nachhaltige und ausgewogene Wachstum. Der MINI ist nicht nur ein Symbol für unsere industrielle Geschichte, sondern steht gleichermaßen auch für die vielversprechende Zukunft unserer Industrie, die wir erreichen wollen.«

Investor Group. Only the MINI brand would remain part of the BMW Group.

In September the new MINI was unveiled at the Paris motor show. The last classic Mini model rolled off the production line at the plant in Longbridge on 4 October 2000. Its production number was 5,387,862. It marked the end of an era spanning 41 years.

2001 In April series production began at the new MINI plant in Oxford.

The British press had initially been critical of BMW's takeover of the Rover Group. The Rover, MG, Land Rover and MINI brands still represented a last bastion of the once world-beating British automotive construction industry. Looking back, the revival of the MINI was a success and can be regarded as one of the major success stories in the history of Anglo-Bavarian collaboration. UK Prime Minister, David Cameron, summed up the major importance of MINI and its association with the production plant in Oxford in June 2011: 'The MINI plant in Oxford has been one of our great manufacturing success stories, they should be hugely proud of their achievements. The company can be proud of the successes it has achieved. They have shown once again that the UK is a major player in the global automotive industry'. Two months later, he added: 'It's a real privilege to drive this true British icon off the production line today. The 2-millionth MINI to be made here in Oxford is a fantastic symbol of the UK's strength in the global automotive industry and a great British manufacturing success story. BMW's continued investment in its UK operations, together with its export success – selling to over 90 countries worldwide – is contributing to a real renaissance for the UK car industry and a brilliant example of the sustainable, balanced growth we are determined to achieve. MINI is not just a symbol of our industrial past, but also the great industrial future we want to build'.

MINI ACV 30 > S./p. 192

Das Autojahr 1997 begann mit einem außergewöhnlichen Ereignis: Im Januar wurde während
der Rallye Monte Carlo eine Konzeptstudie des
MINI Cooper vorgestellt, die ablesen ließ, wie
sich die Marke in Zukunft ausrichten würde. Der
Entwurf verfolgte keine Rückbesinnung mithilfe
eines Retrodesigns, sondern interpretierte das
traditionsreiche Fahrzeugkonzept modern. Die
Zielsetzung war, die Werte des klassischen Mini
mit den Anforderungen an ein modernes Automobil an der Schwelle zum 21. Jahrhundert zu
verbinden. 1997 feierte MINI ein rundes Jubiläum:
Genau dreißig Jahre zuvor hatten die roten
Flitzer aus England die Konkurrenz links liegen
gelassen und bei der Rallye Monte Carlo den
dreifachen Gesamtsieg eingefahren. Dies war
für BMW der gegebene Anlass, die Aufsehen er-
regende Studie zu präsentieren. Eine passende
Bühne als den international vielbeachteten
Rennsportklassiker am Mittelmeer hätte man
nicht finden können.

Der MINI ACV 30 (ACV = Anniversary Concept
Vehicle) beeindruckte durch seinen 160 PS starken
Mittelmotor und das Konzept des Heckantriebs.
Der zweisitzige, rechtsgelenkte Wagen wies ausgesprochen kraftvolle Proportionen als auch
Designmerkmale des klassischen Mini auf – so
den verchromten hexagonalen Kühlergrill und die
großen Rundscheinwerfer. Die vier Zusatzscheinwerfer, die die Frontpartie dominieren, sind eine
unübersehbare Reminiszenz an die Siegerfahr-
zeuge von 1967. Die 16-Zoll-Räder, die extrem
weit in die vier Ecken des Wagens geschoben
wurden, lassen das MINI-typische Gokart-Gefühl
erahnen. Analog zum traditionellen Design ver-

engte sich das Band der Seitenfenster zum Heck
hin, um das farblich in Weiß abgesetzte Dach auf
der Karosserie schweben zu lassen. Die Trapez-
form des klassischen Hatch wurde zusätzlich
gesteigert. Weit ins Dach hineinragende Türen
und eine große Heckklappe unterschieden das
knapp 3,50 Meter lange Zukunftsmobil vom Serien-
modell des klassischen Mini. Deutlicher als das
historische Vorbild betonte das Concept Car von
1997 sein Karosserievolumen über den Rädern
und ließ diese bewusst über die eigentlichen
Grenzen des Wagenkörpers hinaustreten. Damit
gab die Studie ein radikales Designstatement.

Der MINI ACV 30 war das Ergebnis einer er-
folgreichen Zusammenarbeit von Designern aus
dem Hause Rover und BMW. Federführend in
dem Projekt war Adrian van Hooydonk, der Jahre
später die Gesamtverantwortung für das BMW
Group Design übernahm.

/

The year 1997 began with an unusual event: in
January, during the Monte Carlo Rally, a concept
design for the Mini Cooper was unveiled that
set out the brand's direction for the future. Instead of pursuing a return based on a retro design, the study car offered a modern interpretation of the highly traditional car concept. The
aim was to link the values of the classic Mini
with the demands of a modern car on the eve of
the 21st century.

In 1997 Mini celebrated a landmark anniversary: exactly 30 years previously, the red baby
car from England had left the competition trail-
ing behind and had taken triple overall victory at
the Monte Carlo Rally. For BMW, this was the

ideal occasion to present its sensational concept.
There could scarcely be a more fitting stage than
the internationally revered racing sports classic
on the shores of the Mediterranean.

The MINI ACV 30 (ACV = Anniversary Concept Vehicle) cut an impressive figure with its
160 bhp, mid-engined, rear-drive concept car. The
two-seater, right-hand-drive car featured sports-
car proportions as well as the design characteristics inspired by the classic Mini, such as the
chrome-plated hexagonal radiator grille and
the large, round headlights. The four additional
spotlights, which dominate the front section, pay
conspicuous homage to the winning cars from
1967. The 16-inch wheels that have been pushed
far into the four corners of the car are redolent of
the go-kart feeling so characteristic of the MINI.
In a similar way to the traditional design, the
strip of side windows narrowed towards the rear,
making the contrasting white roof seem to float
above the bodywork. The trapezoidal shape of
the classic hatch was also enhanced. Doors
extending far into the roof and a large tailgate
distinguish the just under 3.5-m long car of the
future from the classic serial Mini. More than its
predecessor the 1997 concept car emphasized
the volume of the bodywork through its wheels,
allowing them to extend deliberately beyond the
actual limits of the car's body. As a result, the
study car provided a radical design statement.

The MINI ACV 30 was the product of a successful collaboration between designers from
Rover and BMW. Leading the project was Adrian
van Hooydonk, who years later assumed general
responsibility for BMW Group Design.

Spiritual

Wenige Wochen nach der Präsentation des
MINI ACV 30 stellte die Rover Group ihr neues
Concept Car im März 1997 auf dem Genfer Automobilsalon vor: Der Spiritual war kein neues
Modell der Marke MINI. Vielmehr verbanden
die Hersteller mit ihm ein neues, revolutionäres
Designkonzept. Fast genau vierzig Jahre nach
den ersten Ideenskizzen von Sir Alec Issigonis
präsentierten die Briten ein Fahrzeug, das die
Grundidee seines Begründers transportieren
sollte. Bei der Analyse zukünftiger Herausforderungen standen für die Strategen der Rover
Group die Themen Umwelt und Soziales in einer
spürbar komplexer werdenden Welt im Vordergrund. Die größten Probleme wurden in der zu-
nehmenden Verstädterung, im begrenzten Park-
platzangebot, der Verteuerung der Spritpreise,
der Reduktion des Treibstoffverbrauchs als auch
der Schadstoffe beim Verbrennungsmotor gesehen. In Summe sprachen alle diese Faktoren gegen den allgemeinen Trend zu größeren Autos
und – so Rover – für ein kleineres Fahrzeug.
Wie schon in den Fünfzigerjahren wurde das
folgende Konzept des Spiritual auf die vierköp-
fige Familie zugeschnitten: als Dreitürer mit
einem Platzangebot für vier Erwachsene und dabei doch nur ganze 3,10 Meter lang – womit er
fast genau den Abmessungen des klassischen
Mini entsprach. Der flache Dreizylindermotor
mit 45 kW befand sich unter der Rückbank, der
Tank unter den Vordersitzen, während Ersatz-
reifen, Batterie und Kühler im Frontbereich untergebracht wurden. Mit einem Gewicht von nur
700 Kilogramm, einem Verbrauch von 3 Litern pro
100 km sowie einem Wendekreis von weniger als
9 Metern sowie einer Beschleunigung von 0 auf
100 in 13 Sekunden stand der Spiritual für Effi-
zienz und Wendigkeit. An den alten Mini erinnerten nur noch die runden Kulleraugen.

Neben dem Spiritual bot Rover gleichzeitig
den Spiritual Too an, einen Entwurf für einen
größeren Viertürer, der von einem Vierzylindermotor angetrieben wurde. Dieser Konzeptansatz
sollte das Potenzial einer neuen Modellvielfalt
und »Familienidee« aufzeigen, wie diese bereits
der klassische Mini – ausgehend vom Basis-
modell des Hatch – mit großem Erfolg ausgebildet hatte.

Trotz der zahlreichen Vorzüge und der un-
verminderten Option an Fahrspaß konnten weder
das Konzept des Spiritual noch das nüchterne
One-Box-Design das Fachpublikum und die
Öffentlichkeit überzeugen.

/

A few weeks after the unveiling of the MINI
30, the Rover Group presented its new concept
car at the Geneva motor show in March 1997. The
Spiritual was not a new model from the MINI
brand. Instead, the manufacturers were creating
a new, revolutionary design concept. Almost exactly 40 years after the first ideas outlined by Sir

Alec Issigonis, the British firm unveiled a new design that was intended to communicate the fundamental idea of the original Mini designer. In their analysis of future challenges, strategists at the Rover Group focused on the environment and social awareness in a world that was becoming markedly more complex. The biggest problems were seen to lie in increasing urbanisation, limited parking options, the increase in petrol prices, the reduction of fuel consumption and the reduction of harmful combustion engine emissions. All in all, these factors argued against the general trend towards bigger cars and – according to Rover – in favour of a smaller car. As in the 1950s, the Spiritual's concept was tailored to the family of four: as a three-door model with space for four adults and measuring only 3.10 m long, it had virtually the same dimensions as the classic Mini. The flat, three-cylinder engine delivering 45 kW of power was located under the back seat, the fuel tank was under the front seats, and the spare tyre, battery and radiator were accommodated in the front section. Weighing just 700 kg, with a consumption of 3 litres per 100 km, a turning circle of less than 9 m and acceleration from 0 to 100 km/h in 13 seconds, the Spiritual was the embodiment of efficiency and agility. Only the round headlights provided any reminders of the old Mini.

Alongside the Spiritual, Rover also launched the Spiritual Too, a design for a larger, four-door derivative powered by a four-cylinder engine. This concept approach was intended to highlight the potential offered by a new diversity of models and the 'family idea', as the classic Mini – beginning with the basic hatch model – had done with great success.

Despite the many advantages and unrestricted degree of driving pleasure, the concept of the Spiritual with its sober, one-box design impressed neither the trade audience nor the public.

Der neue MINI / The New MINI

> S. / p. 196

Ein Jahr nach dem Verkauf der Rover Group – die Marke MINI war der BMW Group verblieben – feierte der neue MINI unter dem Dach der BMW Group im Sommer 2001 Weltpremiere und läutete mit den Modellen MINI One und MINI Cooper das Comeback der Marke ein. Die Neuauflage des Klassikers im Kleinwagensegment erwies sich in den folgenden Jahren als überaus erfolgreich.

Das Standardmodell wurde als Zweitürer und vollwertiger Viersitzer angelegt. Mit Frontantrieb, vorn quer eingebautem Vierzylinder-Motor, kurzen Karosserieüberhängen und Platz für vier Insassen wurden die typischen Merkmale des klassischen Mini übernommen. Zwar legten die Außenmaße zu, doch schuf das Design mit den für MINI typischen Proportionen sowie mit unverwechselbaren Gestaltungselementen für Front, Heck und Seitenansicht eine einzigartige Verbindung zwischen alt und neu. Zudem erfüllte der neue MINI einen hohen Anspruch: Als erstes Premium-Fahrzeug im Kleinwagensegment bot er ein großes Maß an Qualität und Sicherheit. Besonderer Wert wurde auf einen möglichst großen Radstand, eine verhältnismäßig breite Spur, straffe Federn und einen niedrigen Schwerpunkt gelegt. Dies führte zu einer extrem guten Straßenlage, die den MINI flink um engste Biegungen kurven lässt und das für MINI typische Gokart-Feeling erzeugt.

Mit dem neuen MINI gelang es den Designern und Technikern, Tradition und Moderne in Balance zu halten. In der Front verleihen große runde Scheinwerfer und der breite Kühlergrill dem MINI ein unverwechselbares Gesicht. Bemerkenswert ist die geschwungene Motorhaube, die nahtlos bis an die Radhäuser reicht und die Frontleuchten integriert. In der Seitenansicht ist auf den ersten Blick die klassische Dreiteilung des Fahrzeugs erkennbar: Sie umfasst den Wagenkörper, ein umlaufendes Band in der Fensterzone und ein Dach, das zu schweben scheint. Die Gürtellinie verläuft von hinten ansteigend und verkürzt damit den Abstand zum Dach im Heckbereich. Elegant und für Kleinwagen ungewöhnlich sind die Türen mit ihren rahmenlosen Fenstern. Ein schönes Detail: die Chromelemente an Kühlergrill, Leuchten und Türgriffen sowie das zwischen Karosserieschulter und Fensterband umlaufende Chromband. Ebenso unverwechselbar sind die schwarzen Kunststoff-Umrahmungen am unteren Bereich der Karosserie, das sogenannte Black Band. Ein extrem kurzes, fast quadratisch geformtes Heck, steil aufsteigende Seitenpartien, verbreiterte Kotflügel und markante Heckleuchten runden das stimmige und moderne Erscheinungsbild des neuen MINI ab.

Im Innenraum fällt sofort das zentrale Element des Tachometers ins Auge, der – mittig platziert – auch die Anzeigen für Tankfüllung und Kühlwassertemperatur enthält. In der Mittelkonsole sind Radio sowie diverse Bedienelemente untergebracht, ebenso der Schalthebel und zwei Cupholder. Der Fond bietet zwei Passagieren bequem Platz. Dank einer teilweise umklappbaren Rücksitzlehne kann man den Kofferraum von 150 auf bis zu 670 Liter Volumen vergrößern. Eindrucksvoll ist das Türdesign mit der Fassung von Armlehne und Ablagefläche. Mit der Karosserie sind auch die Räder gewachsen, die statt auf 10-Zoll- nun auf 15- bis 17-Zoll-Felgen laufen.

Beim neuen MINI kam modernste Technik zum Einsatz: Das Fahrwerk besitzt eine Vorderachse, die auf dem McPherson-Prinzip basiert und damit Vorteile für Gewicht und Raumausnutzung bietet. Gleichlange Achswellen haben einen symmetrischen Rückkopplungseffekt der angetriebenen Vorderräder auf die Lenkung beim Kurvenfahren, Beschleunigen und Bremsen. Eine im Kleinwagensegment einzigartige Multilenker-Hinterachse sorgt für eine optimale Kinematik der Hinterräder. Damit nehmen die Räder stets einen optimalen Winkel gegenüber der Straße ein, und die Reifen haben vollen Kontakt zur Fahrbahn.

Vom hohen technologischen Anspruch zeugen die Scheibenbremsen an allen vier Rädern, das serienmäßige Antiblockiersystem einschließlich einer Kurvenbremskontrolle sowie einer elektronischen Bremskraftverteilung.

Für den Antrieb sorgt ein Vierzylindermotor mit einem Hubraum von 1,6 Litern, 16 Ventilen und Aluminiumzylinderkopf. Während 1959 noch 34 PS genügten, werden ab 2001 vom MINI One 66 kW/90 PS bzw. 85 kW/115 PS beim MINI Cooper erwartet. Bei einem Leergewicht von rund 1100 Kilogramm wird der MINI Cooper zu einem anspruchsvollen Sportwagen.

Eine Spitzenposition im Kleinwagensegment übernahm der MINI auch im Bereich der passiven Sicherheit. Mit einer extrem stabilen Fahrgastzelle, Front- und Seitenairbags sowie optionalen seitlichen Kopfairbags ist der Insassenschutz herausragend. Eine Neuheit ist die serienmäßige Reifenpannenanzeige, die zuvor kein Kleinwagen bieten konnte.

/

One year after the sale of the Rover Group – the MINI brand had remained with the BMW Group – the new MINI celebrated its global première

under the auspices of the BMW Group in the summer of 2001, paving the way for the brand's comeback with the MINI One and MINI Cooper models. The re-make of the classic in the small car sector turned out to be hugely successful in the years that followed.

The standard model was designed as a two-door and fully featured four-door version. The typical features of the classic Mini had been adopted, with front-wheel drive, a transverse four-cylinder engine, short bodywork overhangs and space for four passengers. Although the external dimensions had increased, the design's hallmark MINI proportions and unmistakable design elements for the front, rear and side view created a unique connection between the old and the new. The new MINI also lived up to high expectations: as the first premium vehicle in the small car sector, it offered a high standard of quality and safety. Particular attention was paid to the widest possible wheelbase, a relatively broad track, taut springs and a low centre of gravity. This created excellent road-holding that allows the MINI to corner easily around even the narrowest bends and creates that hallmark MINI go-kart feeling.

With the new MINI, the designers and engineers were able to maintain the perfect balance between tradition and modernity. At the front, large, round headlights and the wide radiator grille gave the MINI an unmistakable 'face'. Particularly striking is the curved bonnet, which extends seamlessly to the wheel arches and integrates the headlights. Viewed side-on, the classic three-part composition of the car is immediately recognizable: it comprises the car body, an all-round strip in the window area and a roof that appears to float. The strap line runs upwards towards the rear, shortening the distance to the roof in the tail section. Stylishly and unusually for small cars, the doors feature frameless windows. Attractive details include the chrome elements on the radiator grille, lights and door handles as well as the chrome strip running between the shoulder of the bodywork and the windows. Equally unmistakable are the black plastic frames on the lower portion of the bodywork, referred to as the 'black band'. An extremely short, almost square-shaped tail, steeply ascending side sections, flared mudguards and striking tail lights complete the coherent and modern appearance of the new MINI.

Inside, the first thing one notices is the central element of the speedometer which, positioned in the middle, also contains the fuel and temperature gauges. The centre console accommodates the radio and various control elements, as well as the gear stick and two cup holders. There is space in the back for two passengers to sit comfortably. Thanks to a partially fold-down rear seat-back, the luggage compartment can be

increased from 150 to a maximum of 670 litres. The design of the doors integrating the arm rest and storage tray is impressive. Along with the bodywork, the wheels have grown too, now running on 15- to 17-inch rather than 10-inch rims.

The new MINI made the most of state-of-the-art technology: the chassis has a front axle based on the McPherson principle, bringing benefits in terms of weight and use of space. Axle shafts of equal length have a symmetrical feedback effect from the powered front wheels on the steering when cornering, accelerating and braking. A multi-arm rear axle, unique in the small car sector, ensures optimum kinematics in the rear wheels. This means that the wheels always adopt the ideal angle relative to the road and that the tyres always have full contact with the road surface.

The disc brakes on all four wheels, ABS system as standard, including cornering brake control, and electronic braking force distribution all bear witness to the MINI's high-tech aspirations.

A four-cylinder 1.6-litre engine with 16 valves and an aluminium cylinder head provide the driving power. Whereas 34 bhp was enough in 1959, the MINI One was expected to deliver 66 kW/90 bhp from 2001 onwards, with the MINI Cooper delivering 85 kW/115 bhp. With an unladen weight of around 1,100 kg, the MINI Cooper is a sophisticated sports car.

The MINI also took the lead in the small car sector in terms of passive safety. With its extremely stable passenger compartment, front and side airbags and optional head airbags, passenger protection is excellent. The standard flat-tyre warning light, which no other small car has been able to offer before, is also new.

MINI Cabrio/Convertible

> S./p. 200

Mit dem Cabrio, das MINI 2004 auf dem Internationalen Automobilsalon in Genf vorstellte, konnte die Palette um ein attraktives Modell erweitert werden. Hatte der klassische Mini gut dreißig Jahre gebraucht, bis er seinen Fans das Fahren unter freiem Himmel ermöglichte, bot das neue MINI Cabrio ein solches Fahrvergnügen weit früher.

Auch das MINI Cabrio ist ein vollwertiger Viersitzer. Seine Frontpartie entspricht dem typischen, sympathischen Gesichtsausdruck der neuen MINI-Ära. Bei geschlossenem Dach fällt das MINI Cabrio zwar etwas flacher aus als die anderen Modellvarianten, doch sind die MINI-typischen Proportionen unverkennbar. Nicht sichtbar ist die speziell beim Cabrio erforderliche Karosseriesteifigkeit. Die steil aufragende Windschutzscheibe wird von der A-Säule eingefasst, in die ein Rohr aus hochfestem Stahl integriert ist. Bei einem Überschlag übernimmt gerade sie eine tragende Rolle. Im Fond schützt ein aus hochfesten Aluminiumrohren gearbeiteter, innovativer Überrollschutzbügel die Fahrgäste.

Eine Schlüsselrolle spielt das hochwertige, strapazierfähige Textilverdeck, das sich auf Knopfdruck elektrohydraulisch bedienen lässt. Zunächst lässt sich der vordere Abschnitt in Form eines integrierten Schiebedachs bis zu 40 Zentimeter zurückfahren. Das vollständige Öffnen und Schließen des Verdecks kann bei einem Tempo von bis zu 30 km/h erfolgen und braucht nicht mehr als 15 Sekunden: Während das Faltdach nach hinten gleitet, werden die Dachholme automatisch eingezogen und gleichzeitig die hinteren Seitenscheiben versenkt. Dank einer praktischen Z-Faltung legt sich das Verdeck kompakt hinter den Rücksitzen ab und macht eine Persenning überflüssig. Das Fehlen einer B-Säule steigert den besonderen Fahrgenuss. Am Heck befindet sich eine nach unten öffnende Gepäckraumklappe mit außen liegenden Scharnieren – ein Detail, das augenzwinkernd ein Konstruktionsmerkmal des ersten klassischen Mini aufgreift. Die Klappe, die als praktische Ladebordwand genutzt werden kann, bietet die Möglichkeit, selbst sperrige Fracht im Wagen unterzubringen.

Eine MINI-typische Innovation in der Welt der Cabrios stellt bei späteren Modellen der Always Open Timer dar. Das links neben dem Drehzahlmesser platzierte Zusatzinstrument hält minutengenau die Fahrzeit fest, in der das Cabrio mit offenem Verdeck gefahren wurde. Aktiviert wird der Timer, sobald der Motor des MINI Cabrio gestartet und das Verdeck vollständig geöffnet ist.
/
The Convertible, which MINI unveiled at the Geneva motor show in 2004, provided the opportunity of adding another attractive model to the portfolio. Whereas it took the classic MINI a good 30 years before its fans could enjoy an open roof, the new MINI Convertible offered this level of driving pleasure much sooner.

The MINI Convertible is another fully featured four-seater. Its front section echoes the characteristic, friendly expression of the new MINI era. With the roof closed, the MINI Convertible looks slightly flatter than the other models, but the hallmark MINI proportions are unmistakeable. The bodywork reinforcement required, especially with the convertible, is invisible. The steeply rising windscreen is enclosed by the A-pillar, into which a tube made of high-density steel is integrated. It literally takes on a supporting role in the event of a roll-over. In the back, an innovative roll-over protection bar made from high-stability aluminium tubing protects passengers.

The high-quality, durable textile roof, which can be operated electro-hydraulically at the touch of a button, plays a key role. To start, the front section in the form of an integrated sliding roof can be moved back up to 40 cm. The soft top can be opened and closed completely at a speed of up to 30 km/h and takes no more than 15 seconds. While the folding roof glides backwards, the roof spars are automatically retracted and the rear side panels simultaneously lowered. Thanks to a practical Z-fold, the soft top is positioned compactly behind the back seats, doing away with the need for a tarpaulin. The lack of a B-pillar makes driving particularly enjoyable.

A downwards-opening luggage compartment with external hinges in the tail is a detail that nods to a design feature of the first classic Mini. The tailgate, which can be used as a handy loading shelf, makes it possible to stow even bulky cargo in the car.

One characteristically MINI innovation in the world of convertibles is the Always Open Timer, available on later models. This additional instrument, positioned on the left next to the speedometer, keeps an accurate record of the time during which the Convertible has been driven with the top down. The timer is activated as soon as the MINI Convertible's engine is on and the roof is completely open.

Eine besondere Vorgeschichte hat dieses Konzeptfahrzeug: 2005 hatte es auf der IAA in Frankfurt seine Weltpremiere gefeiert, als MINI den Weg zu einer neuen Fahrzeugkategorie eröffnete. Das MINI Concept Car griff auf ein Designkonzept zurück, das in den Sechzigerjahren mit den Modellen Morris Mini Traveller und Austin Mini Countryman entwickelt worden war. Nun hieß es, dieses Erbe auf die Gegenwart zu übertragen und neue Möglichkeiten der Karosseriegestaltung und Funktionalität aufzuzeigen. Unter dem Motto »travel the world« wurde die Designstudie in vier Weltstädten mit unterschiedlicher Ausprägung gezeigt, so in Frankfurt, Tokyo, Detroit und schließlich 2006 auf dem Internationalen Automobilsalon in Genf. Gleich zwei historische Anlässe galt es zu feiern: den 100. Geburtstag von Sir Alec Issigonis, dem Schöpfer des klassischen Mini, sowie die Erinnerung an die legendären Motorsporterfolge der Marke vor rund vierzig Jahren. Das MINI Concept Geneva und seine internationale Ausprägung bot hierzu eine zeitgemäße Interpretation.

Mini Traveller und Countryman waren damals der Inbegriff von Vielseitigkeit und Raumangebot. Dieses Volumen im Innenraum musste von außen jedoch gut zugänglich sein. Hier beeindruckte die Studie insbesondere durch ihr innovatives Türkonzept, mit dem die Erschließung und Nutzung des Innenraums völlig neu gedacht wurde. Die zweigeteilte Splitdoor am Heck, die

auf die Türanordnung von Morris Mini Traveller und Austin Mini Countryman zurückgeht, besteht aus Flügeln, die weit außen angeschlagen sind und jeweils nach außen geöffnet werden. Auf diese Weise können sie aus dem Blickfeld der Heckansicht in die Seite geschwenkt werden und bieten einen extrem großzügigen Zugang zum Gepäckraum. Die Konstruktion der aufwendigen Scharniersysteme war eigens entwickelt worden. Diese sogenannte »Parallelogramm-Kinematik« wurde ebenfalls auf die mehr als 1,60 Meter lange Fahrer- bzw. Beifahrertür angewendet.

Das MINI Concept wurde für Genf speziell mit Attributen für den eigenen Renneinsatz bzw. als Begleitfahrzeug ausgestattet. So bietet es einen großen Laderaum, in dessen Boden sich eine Cargobox mit Ersatzteilen versenken lässt. Zwei in klassischem Tartan Red gehaltene Fensterboxen, die Sports Utility Boxes, die in den hinteren Seitenfenstern eingehängt werden, dienen als Werkzeugkisten und halten Gabelschlüssel, Ersatzbirnen und Tripcounter bereit. Im Heckbereich weist der Dachaufbau eine Mulde auf, in der ein Reserverad untergebracht werden kann. An einer Griffleiste, die auch als Spoiler dient, kann dieser Teil des Dachaufbaus heruntergeklappt werden, um das Reserverad mühelos abzunehmen.

Äußerlich wird das Rennsportthema in Form roter Farbakzente aufgegriffen, ansonsten fasziniert die Studie durch einen exklusiven Mix aus

weißen und silbernen Oberflächen mit wenigen Aluminium-Akzenten.

Eine weitere Innovation findet sich im vorderen Innenraum: Hier scheinen die beiden Sitze – nicht mehr als zwei muschelähnliche Schalen – im Raum zu schweben, da sie an ihrer Innenseite über spezielle Tragarme an der vorderen Mittelkonsole verankert sind – eine Konstruktion, die zusätzlichen Fußraum schafft. Das große Zentralinstrument in der Armaturentafel, der Center Speedo, ist drehbar mit Anzeigen auf Vorder- und Rückseite.

Weißes Leder im Innenraum unterstreicht den Eindruck von Modernität, Großzügigkeit und elegantem Ambiente. Wie beim Exterieur kommt auch im Interieur Rot als Akzentfarbe zum Einsatz.
/
This concept vehicle has a very special past. It celebrated its world première at the IAA in Frankfurt in 2005, when MINI was paving the way towards a new category of car. The MINI concept car harked back to a design concept that had been developed in the 1960s with the Morris Mini Traveller and Austin Mini Countryman models. The task now was to transfer this heritage to the present day and highlight new possibilities in terms of bodywork design and functionality. Under the slogan 'travel the world', the design study was showcased in four very different global cities – namely Frankfurt, Tokyo,

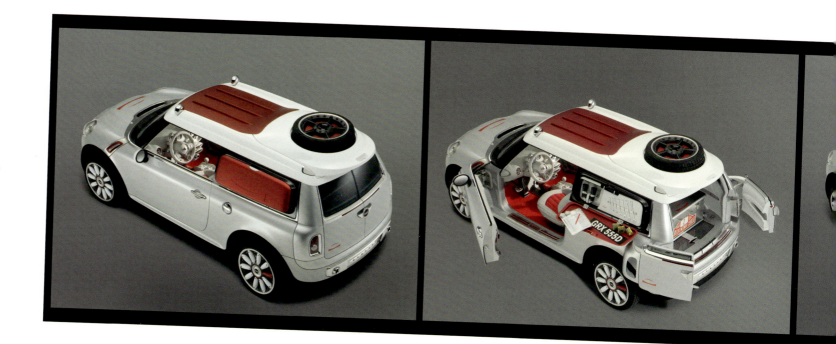

Detroit and finally in 2006 at the Geneva motor show. There were no fewer than two historic occasions to celebrate: the 100th birthday of Sir Alec Issigonis, the creator of the classic Mini, and the commemoration of the brand's legendary motor-sports successes around 40 years previously. The MINI Concept Geneva and its international outlook offered a suitably contemporary interpretation.

At the time the Mini Traveller and Countryman were bywords for versatility and spaciousness. However, this interior volume had to be easily accessible from the outside as well. In this respect the study car was particularly impressive with its innovative door concept, which completely re-imagined access to and use of the interior. The split door at the rear, reminiscent of the door arrangement on the Morris Mini Traveller and Austin Mini Countryman, comprises two wings with wide hinges and which open outwards. This means that they can be swivelled away from the rear field of vision towards the side, and offer extremely generous access to the luggage compartment. The design of the elaborate hinge system had been developed especially for this model. This 'parallelogram kinematics' was also applied to the driver and passenger doors, each measuring more than 1.60 m in length.

For the Geneva show the MINI Concept was specially equipped with features for racing or as an escort vehicle. It offers a spacious loading area, with an integrated cargo box containing spare parts. Two window boxes in classic Tartan Red, the sports utility boxes that are hung in the rear side windows, serve as toolboxes with handy open-ended spanners, spare bulbs and trip counters. At the rear, the roof structure features a depression for stowing a spare wheel. By means of a handle, which also serves as a spoiler, this part of the roof structure can be folded down for easy removal of the spare wheel.

On the outside, the racing theme is continued in the form of red colour highlights. Other fascinating features of the model are its exclusive mixture of white and silver surfaces featuring just a few aluminium highlights.

The front interior houses a further innovation: the two seats – no more than two shell-like structures – appear to float in space as they are anchored on their inner sides to the centre console in front via special supporting arms, thus creating additional foot space. The large central instrument on the dashboard, the centre speedo, can be rotated with displays on both front and reverse.

White leather in the interior emphasizes the impression of modernity, spaciousness and a sophisticated atmosphere. As on the exterior, red is also used in the interior as a highlighting colour.

MINI Clubman / MINI Clubvan

> S. / p. 210

Im Herbst 2007 wurde die Modellpalette um ein innovatives Fahrzeugkonzept erweitert: den MINI Clubman. Sein Name hat Tradition, gab es doch schon 1969 einen klassischen Mini mit dieser Bezeichnung und ähnlicher Charakteristik. Im Vergleich zum MINI One bietet der MINI Clubman eine 24 Zentimeter längere Karosserie und einen um 8 Zentimeter vergrößerten Radstand – und damit größere Ausmaße, die vielfältig genutzt werden können.

Die Seitenansicht ist von einer Keilform geprägt: Während die Dachlinie horizontal verläuft, steigt die Schulterlinie unterhalb des Fensterbandes zum Heck hin leicht an. Von der Front bis zur Windschutzscheibe und der sogenannten A-Säule entspricht der MINI Clubman dem Standardmodell. Erst die verlängerte Karosserie macht ihn einzigartig. Leicht erhöhte Dachflanken – auch als Dune Line bezeichnet – verlaufen über die gesamte Dachlänge hinweg und verleihen der Dachlinie eine schwungvolle Kontur.

Zu den Besonderheiten zählt das Türkonzept: Zusätzlich zur Fahrer- und Beifahrertür bietet der MINI Clubman eine Einstiegsmöglichkeit an der rechten Fahrzeugseite. Diese sogenannte Clubdoor macht Passagieren das Ein- und Aussteigen im Fond leichter, indem sie sich gegenläufig öffnet, also nach hinten ausschwenkt. Mit der zweigeteilten Flügeltür am Heck, im MINI-Jargon auch Splitdoor genannt, wurde ein liebgewonnenes Detail der klassischen Vorfahren – des Morris Mini Traveller und des Austin Mini Countryman – aus den 1960er-Jahren aufgegriffen und neu interpretiert, damit das Be- und Entladen spürbar leichter wird.

Die verlängerte Karosserie bietet auch einen großzügigen Gepäckraum und einen Laderaum mit einem Fassungsvermögen von 260 Litern. Werden die Rücksitzlehnen umgeklappt, erweitert sich das Volumen gar auf 930 Liter – und das bei einem nur 3,96 Meter langen Fahrzeug.

Auf die Basis des MINI Clubman baut der MINI Clubvan auf, der neben zwei Vordersitzen mit einer extrem großen hinteren Ladefläche sowie abgedunkelten Seitenscheiben ausgestattet ist. Das Modell wurde weltweit in einer limitierten Sonderedition von 3000 Fahrzeugen angeboten. Im Rahmen einer Guerilla-Marketingaktion anlässlich der Fußball-Europameisterschaft sorgte ein umgebauter Clubvan 2012 in Polen und in der Ukraine für Furore. Sein Name war Programm: der sogenannte Barbeque-MINI, kurz BBQ MINI, bot mit einer großen Ladefläche die idealen Voraussetzungen für einen ausziehbaren Tisch zum gepflegten Grillen, versehen mit allerlei Equipment, das keine Wünsche offen lässt.

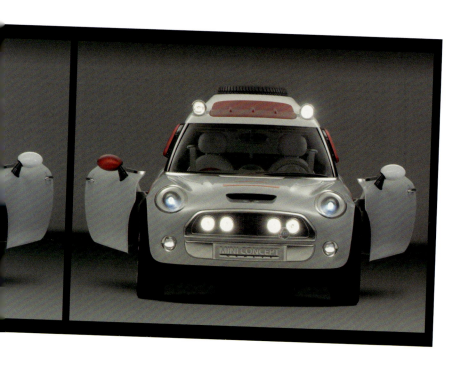

In autumn 2007, the portfolio of models welcomed another innovative vehicle concept – the MINI Clubman. Its name is imbued with tradition, as a classic Mini bearing this name and similar characteristics had appeared back in 1969. Compared to the MINI One, the MINI Clubman featured a chassis that was 24 cm longer and a wheelbase that had been widened by 8 cm, providing larger dimensions that could be used in a myriad different ways.

The side view is characterized by a wedge shape: while the roof line runs horizontally, the shoulder line rises slightly below the windows towards the rear. From the front to the windscreen and the A-pillar, the MINI Clubman is identical to the standard model. Only the elongated chassis makes it unique. Slightly raised roof flanks – also known as the 'dune line' – run the entire length of the roof, bringing a sweeping contour to the roof line.

The door concept is one of the car's highlights. In addition to the driver and passenger door, the MINI Clubman also offers the option of entering the car on the right-hand side. This 'club' door, as it is known, makes it easier for passengers to get in and out of the back of the car by opening in the opposite direction, i.e. swinging out to the rear. The bi-parting wing door at the back, also known in MINI parlance as a 'split' door, was a highly popular detail from the classic predecessors – the Morris Mini Traveller and the Austin Mini Countryman – now revived from the 1960s and reinterpreted to make loading and unloading the car appreciably easier.

The elongated chassis also offers a spacious luggage compartment and a loading area with a capacity of 260 litres. If the rear seat backs are folded down, the volume increases to an impressive 930 litres – and all this from a car that is just 3.96 m long.

The MINI Clubvan was developed from the MINI Clubman, and along with two front seats the vehicle is equipped with an extremely capacious rear loading area and tinted side windows. The model was launched worldwide in an exclusive limited edition of 3,000 vehicles. Again and again MINI has produced spectacular 'guerrilla vehicles', such as on the occasion of the 2012 European Football Championship in Poland and Ukraine. Using ingenuity and imagination a MINI Clubvan was transformed into a showcase Barbeque-MINI (shortened to BBQ MINI). Its tailgate provided the ideal surface for an extendible kitchenette with sophisticated grill, furnished with all kinds of equipment which fulfilled every requirement.

MINI Coupé Concept > S. / p. 214

Die Konzeptstudie eines betont sportlichen Zweisitzers präsentierte MINI erstmals auf der IAA in Frankfurt im September 2009. Gleichzeitig feierte die Marke ihr fünfzigjähriges Jubiläum. Anlass genug, sich auf eine erfolgreiche Tradition zu besinnen und eine Vision für zukünftige Fahrzeugkonzepte und eine prinzipielle Ausrichtung der Modellfamilie zu entwickeln.

Das MINI Coupé Concept vereint eine Vielzahl von Merkmalen, die einen sportlich ambitionierten Fahrstil unterstützen, und entspricht den Anforderungen an ein Coupé mit jeder Menge Fahrspaß: Das Modell bietet zwei Sitzplätze, kompakte Abmessungen, eine ausgewogene Achslastverteilung und einen tieferen Schwerpunkt.

Das Design mit seinen MINI-typischen Merkmalen findet sich auch bei der Konzeptstudie wieder. Die Frontansicht zeigt eine Motorhaube, deren Konturen von großen Radien geprägt sind, Rundscheinwerfer, einen hexogonalen Kühlergrill sowie einen breiten unteren Lufteinlass. Stärker als bei den bereits entwickelten Serienmodellen ist die Windschutzscheibe des Konzeptfahrzeugs geneigt und führt sanft in das Dach über. Die anschließende Dachlinie ist auffallend flach gehalten. Der als Greenhouse bezeichnete, verglaste Teil der Fahrgastzelle fällt deutlich schmaler aus als bei sonstigen MINI-Modellen. Die Seitenansicht bildet beim Coupé eine prä-

gnante Keilform aus. Zum Heck hin mündet das Dach in eine kräftig ausgeformte C-Säule, welche die solide und verwindungssteife Karosseriestruktur betont. Sie lenkt den Blick auf das kurze und flache Heck des Wagens mit seiner angedeuteten Stufenkontur. Hier wird das Aerodynamik-Konzept, welches entscheidend zum Fahrerlebnis des Coupé beiträgt, besonders wirkungsvoll umgesetzt: In Höhe der C-Säule ist ein Heckspoiler angebracht, der das Dach in seiner gesamten Breite überspannt und den aerodynamischen Abtrieb im Bereich der Hinterachse fördert. Ein aktiver Heckspoiler ist zusätzlich in den Gepäckraumdeckel integriert und fährt bei einer Geschwindigkeit von 80 km/h automatisch aus. Gemeinsam mit der geneigten Windschutzscheibe, der auffallend flach gehaltenen Dachlinie sowie einer Karosseriehöhe von nur 1285 Millimetern bietet der Wagen dem Fahrtwind eine geringere Angriffsfläche und reduziert damit nachhaltig den Luftwiderstand.

Kräftige Seitenschweller und spezielle Rohbauversteifungen im Heck haben eine verbesserte Torsionssteifigkeit der Karosserie zur Folge. Durch den Verzicht einer Sitzbank im Fond bietet das MINI Coupé Concept die Möglichkeit, größere Gepäckstücke unterzubringen. Dementsprechend ist das Heck mit einer großen und weit öffnenden Klappe versehen. Ebenfalls weit

öffnende Türen an den Seiten erlauben Fahrer und Beifahrer einen bequemen Einstieg.

Das Interieur ist betont großzügig gestaltet: Horizontale Linien und weit außen platzierte Lüftungsdüsen unterstreichen die Breite des Armaturenbretts. Im Mittelpunkt steht der Center Speedo, ein kreisrundes, MINI-typisches Instrument, das neben der Geschwindigkeitsanzeige gleich mehrere Funktionen und Anzeigen integriert. Zwei Chronoswiss-Uhren zu beiden Seiten des Drehzahlmessers stellen eine kreative Verbindung zu früheren Concept Cars der Marke MINI her und schlagen eine Brücke zwischen dem analogen Zeitalter des klassischen und dem digitalen Zeitalter des neuen MINI: Eine der Uhren ist eine reine Stoppuhr, mit der die Rundenzeiten gemessen werden können, während eine zweite auf klassische Weise die Uhrzeit angibt. /

In September 2009 MINI unveiled the concept study of a distinctly sporty two-seater car at the IAA in Frankfurt. The same year the brand also celebrated its 50th anniversary – reason enough then, to reflect on a successful tradition and develop a vision for future vehicle concepts and outline the direction for the model family of tomorrow.

The MINI Coupé Concept combines a variety of features that contribute to a driving style with

sporty ambitions, and that deliver on the demands made of a coupé with no end of driving pleasure. The model boasts two seats, compact dimensions, a balanced axle distribution and a lower centre of gravity.

The design, with its hallmark MINI characteristics, is also found in the concept study. The front view shows a bonnet whose contours are characterized by large radii, round headlights, a hexagonal radiator grille and a wide bottom air inlet. The windscreen of the concept car is angled more steeply than the previous series models, and transitions smoothly into the roof. The subsequent roof line is much flatter. The glazed section of the passenger compartment, referred to as the 'greenhouse', is considerably narrower than in other MINI models. The side view exhibits a marked wedge shape on the Coupé. Towards the rear, the roof opens out into a powerful C-pillar, which accentuates the solid and warp-free structure of the bodywork. It draws attention to the short, flat rear of the car with its slightly stepped contour. In this instance, the aerodynamics, which contribute much to the Coupé's driving experience, have been executed particularly effectively. At the top of the C-pillar a roof spoiler spans the full width of the roof and promotes aerodynamic downthrust in the rear axle area. An active rear spoiler is also integrated into the luggage compartment cover, extending automatically at speeds of 80 km/h. With the angled windscreen, the strikingly flat roof line and an overall height of just 1,285 mm, the smaller surface of the car offers a reduced frontal area, thus creating a lower drag co-efficient.

Muscular side sills and special shell reinforcements at the rear result in improved torsion stability in the bodywork. The lack of seating in the rear means that the MINI Coupé Concept can carry larger items of luggage. Accordingly, the rear features a large, wide-opening tailgate. Wide-opening doors on the sides make accessing the car a comfortable experience for both driver and passenger.

The interior has a markedly spacious feel. Horizontal lines and widely spaced air vents accentuate the width of the dashboard. In the middle is the centre speedo, a circular, typically MINI instrument which integrates not only the speed indicator but also a number of other functions and displays. Two Chronoswiss clocks on either side of the speedometer forge a creative link to earlier concept cars in the MINI brand and build a bridge between the analogue age of the classic and the digital age of the new MINI: one of the clocks is purely a stopwatch for measuring lap times, while the second displays the time in the classic manner.

Mit dem Countryman erweiterte MINI im Jahr 2010 seine Palette erstmals um einen sportlich ausgelegten Crossover, der Eigenschaften des klassischen Fahrzeugkonzepts des Mini mit einem modernen Sports Activity Vehicle verbindet. Zum ersten Mal ist hier ein MINI optional mit dem Allradantrieb ALL4 ausgestattet, einem Fahrwerk, das Abenteuer jenseits bewährter Straßen erlaubt. Neu sind auch das Konzept eines Viertürers mit großer Heckklappe und eine Karosserielänge jenseits von 4 Metern. Das Erscheinungsbild des MINI Countryman ist vielseitig, robust und von kraftvoller Statur. Das beim österreichischen Produktions- und Entwicklungspartner Magna Steyr in Graz gefertigte Modell fällt durch eine hohe Schulterlinie und eine markante Helmdachform auf. Seine hoch aufbauende Front zeigt einen hexagonal geformten Kühlergrill, der im Vergleich zu bisherigen MINI-Modellen aufrechter steht. Auch die Motorhaube ist markant geformt. Ein nach vorn zulaufendes Powerdome weist auf die Leistungsfähigkeit des darunter liegenden Motors hin. In die Motorhaube eingelassen sind große Scheinwerfer. Für ein kraftvolles Gesamtbild sorgen auch die voluminösen Radhäuser und aufrecht stehende Heckleuchten.

Variabel nutzbar ist der Innenraum, der Platz für bis zu fünf Insassen bietet. Die leicht erhöhte Sitzposition ermöglicht einen bequemen Einstieg und gewährt einen besseren Überblick. Serienmäßig ist der MINI Countryman mit fünf Sitzplätzen und einem auf 1170 Liter ausbaubaren Gepäckraum ausgestattet. Anstelle einer kon-

ventionellen Mittelkonsole bietet MINI eine zwischen dem Fahrer- und Beifahrersitz verlaufende Center Rail an, ein System individuell konfigurierbarer Ablagemöglichkeiten. Mobiltelefone, externe Audiogeräte, Becherhalter und vieles mehr finden auf dieser Bodenachse einen festen Platz im Interieur.

Der optional verfügbare Allradantrieb ALL4 erlaubt eine stufenlose Kraftverteilung zwischen Vorder- und Hinterachse. Für besonderen Fahrkomfort sorgt eine neu entwickelte Fahrwerktechnik, bestehend aus einer Vorderachse mit McPherson-Federbeinen und geschmiedeten Querlenkern, eine Mehrlenker-Hinterachse sowie eine elektromechanische Servolenkung.

Das Konzept des MINI Countryman führt zu einem sportlichen Gesamtpaket und ist weltweit sehr gefragt: Nur zweieinhalb Jahre nach seinem Debut haben sich international bereits 250000 Fahrzeuge verkauft.
/
With the Countryman, MINI extended its portfolio for the first time in 2010 with a sporty crossover vehicle, which combines the properties of the classic Mini car concept with a modern sports activity vehicle. For the first time ever, a MINI can be optionally equipped with the ALL4 all-wheel drive, a chassis that offers adventure off the beaten track. Additional new features are the concept of a four-door car with a large tailgate and an over-4 m chassis length. The MINI Countryman looks versatile, robust and powerful. Produced by Austrian manufacturing and development partner Magna Steyr in Graz, the

model's high shoulder line and striking helmet-shaped roof are remarkable. Its high front exhibits a hexagon-shaped radiator grille which sits more upright compared to previous MINI models. The bonnet too has an unusual shape. A powerdome running towards the front bears witness to the capacity of the engine located underneath. Large headlights are integrated into the bonnet. The voluminous wheel arches and upright-standing tail lights produce a powerful overall image.

The versatile interior offers space for up to five passengers. The slightly raised seating position allows comfortable access and provides a better viewpoint. The MINI Countryman comes with five seats as standard and a luggage compartment that can be expanded to 1,170 litres. Instead of a conventional centre console, MINI provides a centre rail running between the driver and passenger seats with a system of individually configurable storage options. Mobile phones, external audio equipment, cup holders and a whole host of other items can be stowed securely inside the car along this floor axis.

The optional ALL4 all-wheel drive allows continuous power across the front and rear axles. A newly developed chassis technology, comprising a front axle with McPherson spring struts and forged wishbones, a multi-arm rear axle and electromechanical power steering, guarantees an extremely comfortable drive.

The concept of the MINI Countryman produces a complete sporty package that is in great demand worldwide. Just two and a half years after its début, 250,000 cars have been sold.

MINI Beachcomber Concept

Für großes Aufsehen sorgte eine Studie, die MINI im Januar 2010 auf der North American International Auto Show in Detroit vorstellte: Mit dem MINI Beachcomber Concept wurde zugleich eine völlig neue Fahrzeuggattung ins Spiel gebracht, ein Viersitzer mit Allradantrieb und offener Karosserie – maximaler Fahrspaß inbegriffen.

Die Studie kommt ohne Türen, ein konventionelles Dach und abschirmende Karosserieelemente aus. Stattdessen ist sie auf den spontanen und betont aktiven Freizeitgenuss ausgerichtet: Ob trendbewusste Sportler wie Kitesurfer, Wakeboarder oder Triathleten – das MINI Beachcomber Concept fühlt sich dort am wohlsten, wo Strände noch befahrbar sind. Der Name »Beachcomber« kann frei aus dem Englischen mit den Begriffen »Strandgutsammler« oder »Strandwelle« übersetzt werden.

Der Wagen kombiniert einen Allradantrieb mit entsprechend abgestimmter Fahrwerktechnik, mit großzügigen Federwegen, auffallend groß dimensionierten Leichtmetallrädern sowie mit Karosserieelementen, die das Fahrzeug robust erscheinen lassen. In den kraftvoll ausgeprägten Radhäusern sind 17 Zoll große Leichtmetallräder und offroad-taugliche Reifen mit grobem Stollenprofil untergebracht. Auch die massiven Seitenschweller betonen den robusten Gesamtcharakter. Mit diesem Erschinungsbild folgt das MINI Beachcomber Concept den Spuren des Mini Moke, der sich in den Sechziger- und Siebzigerjahren zum Inbegriff von Freizeitvergnügen und Strandaktivitäten entwickelt hatte.

Das moderne Fahrzeugkonzept der Studie des Beachcomber achtet selbst in der Ausführung auf Qualität: Die Karosserie wirkt zwar leicht und offen, doch gelten auch hier Sicherheitsstandards und ein Maximum an Insassenschutz. Die Fahrgastzelle besitzt mit massiv versteiften A-Säulen und einem D-Säulenbügel mit integriertem Querträger einen stabilen Rahmen.

Sollte es plötzlich regnen, wird ein Textilverdeck, das jederzeit einsatzbereit ist, mit wenigen Handgriffen über das Fahrzeug gespannt. Es besteht aus einer extrem leichten und doch robusten Spezialfaser und verfügt an den Seiten sowie am Heck über transparente Kunststoffeinsätze, die als Fenster dienen. Optional bietet MINI passgenaue Kunststoffeinsätze für Dach, Seiten und Heck an.

Vielseitig nutzbar ist der Gepäckraum im Heck. Während am linken Heckelement ein zusätzlicher Staukoffer, der in seiner Form an einen geschlossenen Reserverad-Halter erinnert, befestigt werden kann, ist der rechte Abschnitt dafür ausgelegt, sperrige Gepäckstücke aufzunehmen. Diese können hinten aus dem Fahrzeug herausragen. Mountainbikes oder Wakeboards lassen sich mit Hilfe spezieller Befestigungssysteme sicher verstauen. Auch Surfbretter kann das Fahrzeug problemlos mitführen. Die Aus-

wahl der Farben und Materialien im »Innenraum« ist von der Natur inspiriert. Das Design der Oberflächen und Sitzpolsterungen orientiert sich an den vier Elementen Erde, Feuer, Luft und Wasser. Für besondere Aufmerksamkeit sorgt die Center Rail, ein Element, das aus einer Befestigungsschiene besteht, die von der Armaturentafel bis in den Gepäckraum reicht. Sie kann individuell mit externen Musikplayern, Armauflagen, Aufbewahrungsboxen usw. bestückt werden. Ein integrierter Kabelkanal versorgt eine Kühlbox oder ein Laptop beim Strandabenteuer ausreichend mit Strom.

/

A study car unveiled by MINI in January 2010 at the North American international auto show in Detroit created a huge stir. The MINI Beachcomber Concept represents a completely new type of vehicle: a four-seater car with all-wheel drive and open bodywork with in-built maximum driving pleasure.

The study dispenses with doors, a conventional roof and protective chassis elements. Instead, it is geared towards the spontaneous and supremely active enjoyment of leisure. Whether for fashion-conscious sporty types such as kite surfers, wakeboarders or triathletes, the MINI Beachcomber Concept is in its element on beaches that can be accessed by car. The term 'beachcomber' sums up the attitude of this model perfectly.

The car combines all-wheel drive with co-ordinated chassis technology, generous suspension travel, strikingly well-proportioned alloy wheels and bodywork elements that lend the car a robust appearance. The muscular wheel arches house 17-inch alloy wheels and off-road-compatible tyres with rugged treads. The solid side sills also accentuate the car's sturdy overall look. With this exterior, the MINI Beachcomber Concept is following in the footsteps of the Mini Moke, which became the iconic vehicle for leisure and beach activities in the 1960s and 1970s.

The modern concept of the Beachcomber concentrates on quality even in the execution of the design. The chassis may appear light and open, but safety standards are in force here along with maximum passenger protection. The passenger compartment boasts solidly reinforced A-pillars and a D-pillar bracket with integrated crossbeam to create a more rigid structure.

Should it suddenly start to rain, a textile cover that is always ready for use can be stretched over the car in just a few simple steps. It is made from extremely lightweight yet robust special fibres and has transparent plastic inserts on the sides and the rear which serve as windows. MINI offers optional custom-fit plastic inserts for the roof, sides and rear.

The luggage compartment at the rear offers a wide range of possible uses. While an additional

storage compartment, shaped like a closed spare wheel holder, can be attached to the left rear side, the right-hand section is designed to accommodate bulky items of luggage that can project beyond the rear of the car. Mountain bikes or wakeboards can be securely stowed using specialist fastening systems. The car can even transport surfboards with ease. The choice of colours and materials in the interior has been inspired by nature. The design of the surfaces and seating upholstery draws on the four elements of earth, fire, air and water. The centre rail, a component comprising a fastening rail extending from the dashboard to the luggage compartment, commands particular attention. It can be custom-equipped with external music players, arm rests, storage boxes and so on. An integrated cable duct powers a cool-box or laptop on trips to the beach.

MINI Rocketman Concept

> S. / p. 226

Auf dem Internationalen Automobilsalon in Genf wurde im März 2011 erstmals eine Studie vorgestellt, die in ihren Maßen an den klassischen Mini anknüpft und im Bereich Technologie bereits der Zukunft des Automobilbaus vorgreift.

Mit einer Fahrzeuglänge von nur 3,42 Metern überragt das MINI Rocketman Concept den Vorläufer von 1959 nur um wenige Zentimeter. Maßgeblich war hier weniger die Anbindung an die Tradition, sondern die Vision eines Automobils, das wenig Verkehrsfläche beansprucht, aber viel Innenraum bietet. Bei dem Konzept orientierten sich Designer und Techniker an den elementaren Anforderungen eines modernen mobilen Lifestyles. Gefragt war ein faszinierendes Design in Kombination mit neuen Lösungsansätzen, cleverer Funktionalität und agilem Temperament. Innovativ gibt sich die Studie durch eine puristische, höchst variable Innenausstattung sowie durch Aspekte des konsequenten Leichtbaus. Die Spaceframe-Konstruktion besteht aus Karbon. Deren charakteristische Oberflächenstruktur wird an der Front, im Türbereich sowie im Innenraum partiell sichtbar. Damit gibt die Studie zu erkennen, dass sie auf ein Leichtbaukonzept mit geringem Spritverbrauch abzielt.

Die Grundformen werden von gerade verlaufenden Linien bestimmt, die Flächen sind straff ausmodelliert. Großformatige, runde Scheinwerfer und der von einem Chromrahmen eingefasste Kühlergrill bestimmen die Front. Die Struktur der Lichtquellen wurde weiterentwickelt: Zentral angeordnete LED-Einheiten steuern das Fernlicht und werden von einem markanten Leuchtring umgeben, der das Abblendlicht er-

zeugt. Die Heckleuchten sind als trapezförmige Bügel ausgebildet, die sämtliche Lichtfunktionen integrieren. Mit einer außergewöhnlichen Lichtinszenierung wartet auch der Dachbereich auf: Das Glasdach wird durch beleuchtbare Streben unterteilt, die die Form des britischen Union Jack symbolisieren. In unbeleuchtetem Zustand strahlen die Streben in hellem Weiß. Bei Dunkelheit sorgen die integrierten Lichtleiter für ein effektvolles Nachtdesign. Es gibt auch noch eine Steigerung: Auf Höhe der Brüstungslinie, also der Stelle, an der beim regulären MINI ein Chromrahmen verläuft, besitzt das MINI Rocketman Concept eine Lichtleiste.

Einen besonderen Anteil an der Innovationsfülle haben die beiden seitlichen Türen. Beim Öffnen schwenken sie samt Türschweller nach außen, sodass die Passagiere bequem einsteigen und Platz nehmen können. Die vorn angesetzten Türen verfügen über ein Doppelscharniergelenk, das selbst auf begrenztem Raum einen großen Öffnungswinkel ermöglicht. Auch bei geschlossenen Türen ist dieses Element gut erkennbar. Die zweigeteilte Heckklappe hat ein ebenso außergewöhnliches Konzept: Sie besteht aus einem am Dach angesetzten Segment, das beim Öffnen weit nach oben schwingt. Der untere Abschnitt ist in Form eines Schubfachs ausgebildet, das bis zu 35 Zentimeter aus der Karosserie ausfahren und Gepäck oder Reiseutensilien aufnehmen kann.

Im Inneren geht das Design dieser Studie ebenfalls neue Wege: Hier werden Lederbezüge und hochglanzlackierte Oberflächen mit Armauflagen und Dekorleisten aus einem gepressten Spezialpapier kombiniert, wobei die Leisten mit einem LED-Lichtleiter hinterleuchtet werden.

Ein neuartiges Bedienkonzept orientiert sich an den vielschichtigen Anforderungen moderner Zielgruppen. Die zentrale Bedieneinheit kann entnommen und am Computer konfiguriert werden. Alle Elemente sind auf das Lenkrad konzentriert. Zusätzlich ist in die rechte Lenkradspeiche ein Trackball eingelassen, ein Element, das aus der Welt der Computerbedienung abgeleitet wurde.

/

At the Geneva motor show in March 2011, MINI unveiled a study car that echoes the classic Mini in its dimensions and anticipates the future of automotive construction in terms of its technology.

With a vehicle length of just 3.42 m, the MINI Rocketman Concept is just a few centimetres longer than its 1959 predecessor. The connection with tradition was less important in this context, but instead it was the vision of a car that takes up less road space and offers plenty of interior room. With this concept, the designers and engineers focused on the fundamental requirements of a modern mobile lifestyle. A compelling design combined with innovative solutions, smart functionality and an agile disposition was needed. The study car demonstrates its innovation through purist, highly versatile interior fittings and a rigorously lightweight construction. The spaceframe construction is made of carbon. Its characteristic surface structure is partially visible at the front in both the door area and the interior. As a result, the study shows that its aim is to create a lightweight construction with low fuel consumption.

The basic shapes are determined by straight lines, while the surfaces are tightly sculpted. Large-format, round headlights and the chrome-framed radiator grille dominate the front. The structure of the lights has been refined: centrally arranged LED units direct the high beam and are surrounded by a striking ring of lights that produces the low beam. The tail lights are designed

as trapezoidal stirrups that integrate all the light functions. The roof area also has an unusual lighting set-up: the glass roof is divided into sections by illuminated braces that symbolise the outline of the Union Flag of the United Kingdom. Unilluminated, the braces glow bright white. In the dark, the integrated optical fibres produce an effective night-time design. And as another highlight, at the level of the sill line, where a chrome frame usually appears on the normal MINI, the MINI Rocketman Concept boasts an illuminated strip.

The two side doors comprise a particularly strong element in the wealth of innovations that make up this car. When opened, both they and their door sills open outwards, allowing passengers to enter and take a seat comfortably. The front doors have a double-hinge joint that allows them to be opened wide even in confined spaces. This element is clearly recognizable even with the doors closed. The two-part tailgate also boasts an unusual concept: it comprises a section attached to the roof which swings up when opened. The lower section is in the form of a drawer that can be extended up to 35 cm beyond the bodywork to accommodate luggage or travel accessories.

Inside, the design of this study car also breaks new ground. Leather covers and high-gloss painted surfaces combine with arm rests and trim strips made from a special paper moulded into shape; the trim strips are back-lit with LED fibre-optics.

An innovative operating concept is aimed at the many and varied needs of modern target groups. The central control unit can be removed and re-configured on the computer. All of the elements are focused on the steering wheel. The right-hand steering wheel spoke also has an integrated track ball – another element borrowed from the world of computing.

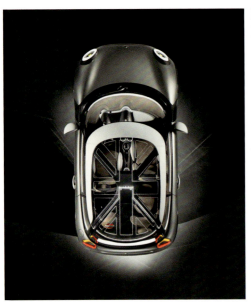

MINI Roadster > S. / p. 222

Der kompakte Roadster ist der erste offene Zweisitzer in der Geschichte des MINI, das einzige Modell ohne klassisches Pendant in der Markenhistorie.

Im Vergleich zum Cabrio ist der Roadster mit seiner stärker geneigten A-Säule, der kontinuierlich leicht ansteigenden Brüstungslinie und dem flach auslaufenden Heck deutlich mehr gestreckt. Selbst bei geschlossenem Verdeck unterscheidet sich der Roadster durch eine elegant gezogene Silhouette. Auch fällt die Fahrzeughöhe 20 Millimeter geringer aus als beim Cabrio.

Die ausgeprägte Neigung der A-Säule und der Frontscheibe reduziert die Stirnfläche der Karosserie und damit den Luftwiderstand. Außerdem ist der Roadster mit einem sogenannten »aktiven« Heckspoiler ausgestattet. Das in die Gepäckraumklappe integrierte Element fährt automatisch aus, sobald der Roadster eine Geschwindigkeit von 80 km/h erreicht. Ein besonderes Merkmal sind die aus Edelstahl gefertigten, an ihren Vorderseiten gepolsterten Überrollbügel. Gemeinsam mit dem verstärkten Rahmen der Windschutzscheibe bieten sie im Fall eines Überschlags den erforderlichen Insassenschutz.

Puristische Sportlichkeit prägt den Wagen. So ist die Fahrgastzelle auf zwei Sitzplätze beschränkt. Das klassische Textilverdeck lässt sich zwar schnell und einfach, aber eben nur manuell öffnen und schließen. Es ist robust und uneingeschränkt das ganze Jahr über nutzbar. Ein semi-automatisches Verdeck kann inzwischen als Sonderausgabe bestellt werden.

2012 wurde einer der MINI Roadster dank einer extravaganten Dekoration zum begehrten Sondermodell. Anlass war das zwanzigjährige Bestehen des Life Ball in Wien, einer vielbeachteten Charity-Veranstaltung. Dort wurde der Roadster im Rahmen der AIDS Solidarity Gala versteigert. Farbwahl und Motive tragen die künstlerische Handschrift von Franca Sozzani, Chefredakteurin der italienischen *Vogue*. Inspiriert durch die Vorstellung schöner, eleganter Damen, die in der Vergangenheit mit einem leichten, um den Kopf drapierten Schal Cabrio fuhren, ließ Sozzani den Wagen mit einem Deep-Purple-Matt-Lack sowie goldglänzenden Rallye-Streifen überziehen und dekorierte die Oberflächen anschließend mit filigranen Blüten. /

The compact Roadster is the first open two-seater in the MINI's history, and the only model without a classic equivalent in the story of the brand.

Compared to the Convertible, the Roadster's more aggressive A-pillar, continuously gently rising sill line and flat tapered rear lend it a significantly more elongated appearance. Even with the top closed, the Roadster's elegant silhouette stands out. The height of the car is also 20 mm lower than the Convertible.

The marked inclination of the A-pillar and the front windscreen reduce the front face of the bodywork and thus its wind resistance. The Roadster is also equipped with an 'active' rear spoiler. Integrated into the luggage compartment flap, the spoiler extends automatically as soon as the Roadster reaches a speed of 80 km/h. One particular feature is the roll-over bars, made from stainless steel and padded on the front. Together with the reinforced frame of the windscreen, they offer the required amount of passenger protection in the event of a roll-over.

The car is characterized by pure sportiness. As a result, the passenger compartment is restricted to two seats. The classic textile soft top can be opened and closed quickly and easily, but only manually. It is sturdy and can be used unrestrictedly all year round. A semi-automatic roof can now be ordered as a special edition.

In 2012, one of the MINI Roadsters became a much sought-after special model thanks to its extravagant trim. It was on the occasion of the 20th anniversary of the Life Ball in Vienna, a highly prestigious charity event. The Roadster was auctioned as part of the AIDS Solidarity Gala. The choice of colour and motifs bore the artistic signature of Franca Sozzani, Editor-in-Chief of Italian *Vogue*. Inspired by the idea of beautiful, stylish women who, in the past, would drive a convertible with a flimsy scarf wrapped around their heads, Sozzani covered the car with a deep purple matt paint and shimmering gold rally stripes, then decorated the surfaces with delicate flowers.

MINI Paceman > S. / p. 230

Mit dem großen Erfolg des MINI Countryman reifte der Plan, ihm ein Coupé folgen zu lassen, das erste Sports Activity Coupé im Segment der Klein- und Kompaktwagen. Wie der MINI Countryman wird auch der MINI Paceman im österreichischen Graz gefertigt und verfügt wie sein Kollege optional über Allradantrieb ALL4. Ebenso stehen verschiedene, betont kraftvolle Motoren zur Auswahl. Das Angebot des MINI Paceman reicht vom 112 PS starken Cooper D bis zum sportlichen Cooper S mit 190 PS. Selbst das Topmodell mit seinem 140 kW/190 PS starken und 1,6 Liter großen Vierzylinder-Turbobenziner verbraucht trotz beeindruckender maximaler Fahrleistung von 218 km/h und einer Beschleunigung von Tempo 0 auf 100 in 7,5 Sekunden nur 6,0 Liter. Neben dem Sportfahrwerk mit einer Absenkung um 11 Millimeter, das zur Serienausstattung zählt, unterstreichen auch 19 Zoll große Räder den sportlichen und athletischen Gesamteindruck.

Im Vergleich zum MINI Countryman gibt es bei diesem Modell ein paar Abweichungen, die dem Umbau zu einem Coupé zu verdanken sind: Die Karosserie ist gestreckt, die Dachlinie fällt sanft nach hinten ab, und die zwei großen, langen Coupétüren fließen über eine breite Schulterpartie zum Heck, das zwei großformatige, markante Rückleuchten prägen. Wie schon das große Vorbild, der Countryman, trägt der MINI Paceman seinen Modellnamen selbstbewusst in großen Chromlettern auf der Heckklappe.

Eine weit nach oben schwingende Heckklappe und eine niedrige Ladekante erlauben ein komfortables Be- und Entladen. Als geräumig und variabel nutzbar erweist sich das Gepäckabteil, das sich von 330 auf 1080 Liter erweitern lässt. Der Innenraum beeindruckt mit großzügigen Platzverhältnissen und vier komfortabel ausgeformten Sportsitzen aus Leder, darunter zwei bequemen Leder-Lounge-Sitzen im Fond.

Wie der MINI Countryman bietet auch der MINI Paceman die eigens von MINI entwickelte Center Rail, in die Ablageboxen für Smartphone, Sonnenbrillen oder Becherhalter eingeklinkt werden können.

/

In light of the MINI Countryman's massive success, plans were made to introduce a coupé – the first sports activity coupé in the compact and sub-compact car sector. Like the MINI Countryman, the MINI Paceman is also produced in Graz in Austria and can optionally be delivered with the ALL4 four-wheel drive. The series also offers a variety of engines with special focus on performance. The MINI Paceman series includes various models – from the 112 bhp Cooper D to the sporty Cooper S, featuring 190 bhp. Even the top model, equipped with a 140 kW/190 bhp 1.6-litre four-cylinder turbo petrol engine, offers an efficiently low consumption of only 6.0 litres in spite of an impressive top speed of 218 km/h and acceleration from 0 to 100 km/h in 7.5 seconds. In addition to the standard sports car chassis lowered by 11 mm, 19-inch wheels emphasize its sporty and athletic appearance.

Compared with the MINI Countryman, upon which it is based, there are many differences which arose in the conversion to a coupé. The body is elongated, the roofline slopes down gently towards the rear and two large and long coupé doors seem to flow towards the broad shoulder section at the rear, which is characterized by two large tail lights. Like the Countryman, the MINI Paceman confidently bears its model name in big chrome letters on the tailgate.

A tailgate that opens high and a low boot sill make loading and unloading the boot easy. The spacious and flexible luggage compartment can be expanded from 330 to 1,080 litres. The passenger compartment boasts an impressively spacious layout and four comfortably designed

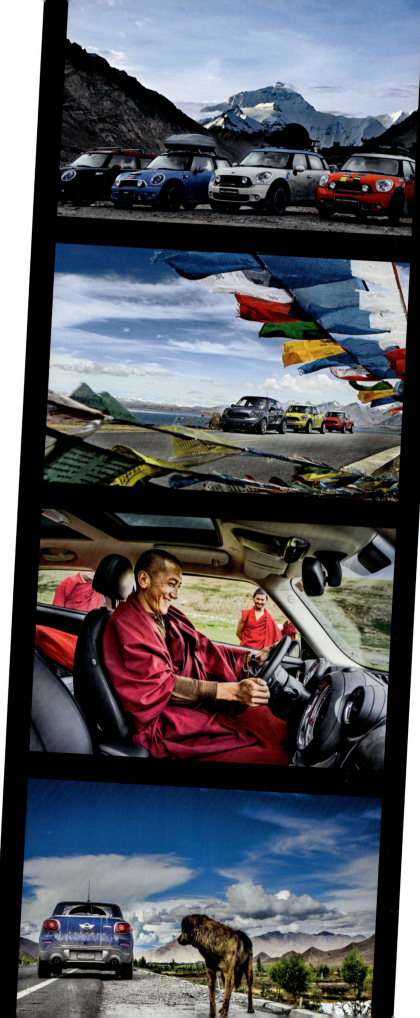

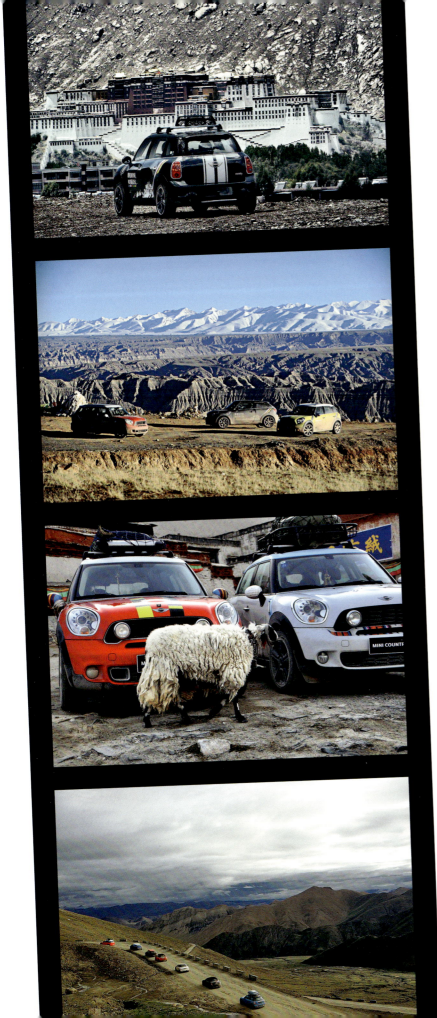

leather sports seats, including two convenient leather lounge seats in the back.

Just like the MINI Countryman, the MINI Paceman features the special centre rail developed by MINI, to which storage compartments for smartphones, sunglasses or cup holders can be attached.

MINI goes to Tibet

Im Sommer 2013 erfolgte in Fernost eine der außergewöhnlichsten Reisen, bei denen der MINI Paceman seine Crossover-Eigenschaften unter Beweis stellen konnte: Der fast siebzigjährige Fotograf Demei Cheng hatte im Laufe seines Lebens zahlreiche Reisen durch die öde, unermessliche Weite von Tibet unternommen und dabei stets unterschiedliche Routen gewählt. Begonnen hatten diese Abenteuer in den späten Sechzigerjahren mit einer wissenschaftlichen Expedition an der Seite seines Vaters. Anfang der Neunzigerjahre nahm jener wiederum seinen Sohn auf die langen Erkundungsfahrten mit. So war eine ganze Familie über Jahre mit der einzigartigen Landschaft Asiens eng verbunden. Ungeachtet der unzähligen Fahrten und Wanderungen bis zum »Dach der Welt« fehlte Demei schließlich nur noch die schwierigste der fünf möglichen Routen durch Tibet. Am 6. August 2013 war es dann soweit: An fünf verschiedenen Punkten in Tibet machten sich MINI-Fahrzeuge auf den Weg. Ihr Ziel: der Mount Everest. In den folgenden Wochen legten sie gemeinsam durch unwegsame Wüsten und über steile, staubige Bergstraßen rund 70000 Kilometer zurück.

Mit dabei waren Demei Cheng und seine Frau, die sich für die äußerst anspruchsvolle Strecke Xinjiang–Tibet entschieden hatten und nun mit ihrem Sohn – einem erfahrenen Tibetreisenden – und ihrem dreizehnjährigen Enkel im MINI Paceman Platz fanden. Ausgerechnet diese Route gilt als eine der schwersten und kältesten der Welt. Sie führt von Kargilik in Xinjiang vorbei an zehn Gipfeln der Kunlun-Berge mit einem atemberaubenden Panorama und mitten durch die Gletscher in Aksai Chin bis nach Tingri. Auf der Strecke von knapp 1000 Kilometern durch endlos weites und bisweilen menschenleeres Gebiet wurden Höhenunterschiede von 1300 Metern über dem Meeresspiegel bis auf 5000 Meter überwunden. Der MINI Paceman meisterte die extreme Beanspruchung ohne Probleme. Auch das Ehepaar Cheng blieb bei guter Gesundheit und fand sogar Zeit, am höchsten Punkt der Strecke, dem Gipfel der Guge-Berge, anzuhalten und den 67. Geburtstag der Ehefrau zu feiern. Anfang September 2013 trafen die MINI, die an fünf verschiedenen Punkten gestartet waren, am sogenannten Lhasa Eye zusammen, um gemeinsam zur letzten Etappe aufzubrechen. Auf

5248 Meter Höhe erreichte der Track den Tempel Rongbu, den höchstgelegenen Buddhisten-Tempel der Welt, und schließlich – nach insgesamt 70000 Kilometer Fahrt durch extreme Witterung und herausforderndes Gelände – das Basislager des Mount Everest. Einer der außergewöhnlichsten Familienausflüge fand damit sein glückliches Ende.

/

In summer 2013, one of the most unusual journeys took place in the Far East, during which the MINI Paceman demonstrated its crossover abilities. Over the course of his life, photographer Demei Cheng, who is almost 70 years old, has made many journeys through the barren vastness of Tibet, always choosing a different route. His adventures started in the late 1960s on a scientific expedition with his father. In the early 1990s Demei took his own son on long journeys of exploration. Over the years, the whole family became closely connected with the unique Asian landscape. Despite his innumerable trips and tours to the 'roof of the world', Demei had still not taken the most difficult of the five routes that cross Tibet. This tour began on 6 August 2013 and the MINI cars started off from five different points in Tibet. Their destination was Mount Everest. During the weeks that followed together they drove some 70,000 km through rough deserts, and steep, dusty mountain roads.

Among the travellers were Demei Cheng and his wife who had both decided to go on this extremely arduous route from Xinjiang to Tibet. Now they found themselves together with their son – an experienced Tibet traveller – and their 13-year-old grandchild in a MINI Paceman. This route in particular is regarded as one of the most difficult and coldest in the world. It runs from Kargilik in Xinjiang past ten peaks of the Kunlun Mountains with a breathtaking panoramic view, right through the glaciers in Aksai Chin to Tingri. They surmounted changes in altitude between 1,300 and 5,000 m above sea level on a route of nearly 1,000 km through endless and sometimes deserted landscapes. The MINI Paceman tackled the extreme conditions without any problems. The Chengs too remained fit and healthy and even found time to stop at the highest point along the route, the peak of the Guge Mountains, to celebrate Mrs Cheng's 67th birthday. At the beginning of September 2013, the MINIs, which had started from five different points, met up at a place known as the Lhasa Eye for the final leg. At an altitude of 5,248 m, the track reached the temple of Rongbu, the highest Buddhist temple in the world, and finally – after a journey totalling 70,000 km through extreme weather and challenging terrain – they arrived at Mount Everest base camp. And so one of the most unusual family outings had a happy ending.

MINI XXL

Der gewöhnliche MINI – der klassische wie der neue – lebt von bestimmten Proportionen und einer Maximallänge von 3,05 bzw. 3,62 Metern.

Am 11. August 2004 aber gab es in der Geschichte von MINI einen Höhepunkt: Während der Olympischen Sommerspiele 2004 in Athen wurde während der Veranstaltung des MINI@ Galea Club ein Fahrzeug in spektakulärer Übergröße vorgestellt und von der internationalen Presse, Promis und MINI-Fans mit Begeisterung aufgenommen. Der kleine Kompaktwagen im unverwechselbaren MINI-Design war auf die Maße einer 6,30 Meter langen Stretch-Limousine gestreckt worden und mutierte als MINI XXL zu einer Aufsehen erregenden Prachtkutsche – und das mitten im Klima sportlicher Höchstleistungen. Im fernen Los Angeles, wo man über ein profundes Know-how im Bereich Stretch-Limousinen verfügte, hatte man den ursprünglichen MINI Cooper S im roten Dress zum Viertürer umgebaut und die Karosserie derart ge-

längt, dass sie von einer dritten Hinterachse gestützt werden musste. Für eine leistungsstärkere Motorisierung sorgte ein John-Cooper-Works-Tuning-Kit.

Der bequeme Passagierfond erschien in schwarzem Lederinterieur. Er bot vier Personen jede Menge Komfort und Unterhaltung, darunter einen hochfahrbaren Flatscreen-Fernseher, der die Fahrerkabine vollständig von den Gästen abtrennte, DVD Player, Radio/CD, Klimaanlage, Schiebedach und Telefon – schließlich mussten »Intra-Limo«-Gespräche mit dem Chauffeur während der Fahrt möglich sein. Die größte Überraschung fand sich jedoch im hinteren Bereich: Dort prunkte ein waschechter, abdeckbarer Whirlpool auf dem Heck. Das Badevergnügen war jedoch nur im Stillstand erlaubt.
/
Both the classic and the new regular MINI alike are based on certain proportions and a maximum length of 3.05 and 3.62 m respectively.

A very special day in MINI history occurred on 11 August 2004. A spectacularly oversized car was presented at an event of the MINI@Galea Club during the Summer Olympic Games of 2004 in Athens, and was received enthusiastically by the international press, celebrities and MINI fans. The small compact car with its distinctive design had been elongated into a 6.30 m stretch limousine, transformed into a magnificent, eye-catching MINI XXL – right in the middle of a top sporting event. Stretch-limousine experts in Los Angeles had re-built the original red MINI Cooper S into a four-door car and extended the car body so far that it had to be supported by a third rear axle. A John Cooper Works Tuning Kit was used to increase engine power.

The comfortable rear passenger compartment was upholstered in black leather. For four people the car offered ample comfort and entertainment, including a retractable flat screen TV completely separating the driver's compartment from the

passengers, a DVD player, radio/CD, air-conditioning, a sunroof and a phone for 'intra-limo' conversations with the driver during the journey. The rear section, however, offered the most amazing surprise: a real, coverable whirlpool was integrated into the back, although bathing was only allowed when the MINI XXL was parked.

MINI Superleggera™ Vision

Ein ganz besonderes Concept Car ist der MINI Superleggera™ Vision, der im Rahmen des Concorso d'Eleganza Villa d'Este im Mai 2014 erstmals der Öffentlichkeit vorgestellt wurde.

Den durch einen Elektromotor angetriebenen, offenen Zweisitzer schuf MINI gemeinsam mit Touring Superleggera, dem seit 1926 bestehenden Design- und Karosseriehersteller aus Mailand. Es war das erste Mal, dass die britische Automarke mit einer italienischen Designfirma zusammenarbeitete. Die für die Marke MINI typischen Designmerkmale wurden bei diesem Modell mit der sportlichen Eleganz und Handwerkskunst der italienischen Carrozzeria in Verbindung gebracht, coole Britishness und italienisches Flair zu einem emotionalen Konzept vereint. Das Ergebnis ist ein klassischer, sehr kompakt erscheinender Roadster, der besonderen Fahrspaß verspricht. Die Proportionen des MINI Superleggera™ Vision sind perfekt ausbalanciert. Die Motorhaube ist extrem gestreckt, der Radstand sehr lang, die Überhänge entsprechend kurz. Die Front mit den beiden großen Scheinwerfern und dem sechseckigen Grill weisen die typischen Designelemente des MINI auf. Die für die Marke charakteristischen »Bonnet Stripes« sind dreidimensional in die Motorhaube eingeprägt und münden in polierte Aluminiumelemente. Die Seiten sind geprägt von der typischen »Touring-Linie« des norditalienischen Traditionsbetriebes. Umgeben von straffen, fugenlosen Flächen zeichnet sie eine präzise verlaufende, spannungsvolle Bewegung nach. Das Heck ist flach und breit angelegt und zeigt zwei besondere LED-Heckleuchten in Form eines zweigeteilten Union Jack.

In aufwendiger Handarbeit wurde ein Einzelstück gefertigt, das die Tradition des klassischen Karosseriebaus fortführt. Sämtliche Aluminiumbleche waren in relativ großen Teilen von Hand geformt worden. »Was mit dem classic Mini vor 55 Jahren begann, führt das MINI Superleggera™ Vision elegant fort: die Reduktion auf das Wesentliche. Sein minimalistisches, aber gleichzeitig hochemotionales Design verkörpert die dynamische Essenz eines MINI. Gleichzeitig vereint es in sich Herkunft und Zukunft der Automobilindustrie, also traditionelle Karosseriebaukunst und moderne Designsprache, in einzigartiger Schönheit.«, beschrieb Anders Warming, der Leiter des Bereichs MINI Design, das Unterfangen der beiden Unternehmen.

Das Resultat stellt in jeder Hinsicht zufrieden: Der Wagen zeichnet sich durch ein hochwertiges, geschlossenes und elegantes Erscheinungsbild aus, nicht zuletzt wegen der Lackierung in Como Blue, einem klassischen Blaumetallic mit moderner Anmutung. Front- und Seitenteile aus Karbon unterstreichen den sportlichen Charakter des Konzeptfahrzeugs, dessen Innenraum mit puren Materialien wie Leder, Aluminium und Schwarzchrom veredelt wurde. Das Armaturenbrett besteht aus einem einzigen Aluminiumblech und verweist ebenso wie die Türen und das Dreispeichen-Lenkrad auf eine traditionelle Bauweise. Das Center-Instrument in der Mitte der Instrumententafel orientiert sich an der Formensprache von MINI.
/
The MINI Superleggera™ Vision is a very special concept car, and was introduced to the public during the Concorso d'Eleganza Villa d'Este in May 2014.

MINI created the electrically powered, open two-seater in co-operation with Touring Superleggera, the design and car body manufacturer from Milan, established in 1926. This was MINI's first co-operation with an Italian design company. In this model, the distinctive MINI design features were combined with the Italian car body manufacturer's sporty elegance and craftsmanship, uniting British cool and Italian flair to create emotional appeal. The result was a classic roadster with a highly compact appearance, offering driving pleasure at its best. The MINI Superleggera™ Vision features perfectly balanced proportions. The bonnet clamshell is highly elongated, the wheelbase very long and the body overhang correspondingly short. The front section with its big headlights and the hexagonal radiator grille displays typical MINI design features. The brand's typical bonnet stripes are embossed three-dimensionally into the bonnet and end in polished aluminium accents. The sides are characterized by the 'Touring line', typical of the traditional North Italian company. Surrounded by taut, seamless surfaces, it creates precise, vibrant movement from front

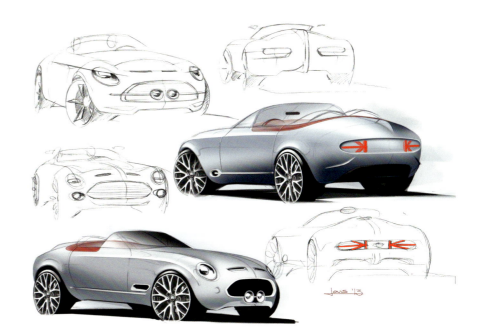

MINI Clubman Concept

to rear. The rear is flat and broad, including two special LED tail lights that form a two-part Union Flag.

Elaborate craftsmanship has gone into creating a unique specimen that continues the tradition of classic bodywork. All the relatively large-format aluminium sheets were shaped by hand. Anders Warming, Head of MINI Design, described the two companies' initiative: 'The MINI Superleggera™ Vision elegantly perpetuates what the Classic Mini started 55 years ago: reduction to the essentials. Its energetic, minimalistic design embodies the dynamic essence of an automobile. At the same time it creates unique emotional beauty in combining the past and future of the automotive industry, i.e. traditional coachwork craftsmanship and modern design styling'.

In every respect the result is perfect. The car boasts a high-quality, compact and elegant appearance, not least due to the Como Blue paint finish, a classic metallic colour with a modern feel. Front and lateral carbon parts emphasize the concept car's sporty character, while the interior was refined using pure materials, such as leather, aluminium and black chrome. The dashboard consists of a single sheet of aluminium, highlighting traditional car building methods, just like the doors and the three-spoke steering wheel. The Centre Instrument in the middle of the dashboard centre draws on MINI design language.

Auf dem Genfer Autosalon 2014 präsentierte MINI das Clubman Concept, ein elegantes, in warmem Berry Red und Blade – einem metallischen Grau – gehaltenes Fahrzeug mit 4,22 Meter Länge und 1,84 Meter Breite. Mit diesen Ausmaßen übertraf es den aktuellen MINI Clubman. Hatte sich der erste Mini von 1959 vor allem dadurch ausgezeichnet, dass er maximalen Raum auf minimaler Fläche bot, so wurde dieser Anspruch nun auf eine neue – größere – Fahrzeugklasse übertragen.

Mit dem Clubman Concept stieg MINI erstmals in ein Segment ein, das sich primär durch Funktion und Nutzen definiert. Ein außergewöhnliches Design, besondere Farb- und Materialkombinationen und zahlreiche clevere Ideen treffen hier auf die MINI-typischen Designmerkmale im Exterieur. Hervorzuheben ist die langgestreckte Dachlinie, welche den großen Nutzraum des Fahrzeugs andeutet, vier Türen, großzügige seitliche Flächenverläufe und ein breites Heck. Auch die charakteristischen Splitdoors im Heck wurden beibehalten, jedoch durch das Band der Fenstergrafik akzentuiert. Gemeinsam mit den horizontal angeordneten Heckleuchten sorgt es für einen modernen Gesamteindruck. Auch verfügt das MINI Clubman Concept über zahlreiche aerodynamische Maßnahmen wie Dachspoiler, Air Breather und Air Curtain, wobei sämtliche Öffnungen und Auslässe in den Fahrzeugkörper integriert sind.

Das Interieur zeichnet sich durch äußerst hochwertige Materialien aus. Vor allem die

Ellipse als gestaltendes Grundelement zieht alle Aufmerksamkeit auf sich. Zusammen mit kostbarer Holz- und Lederausstattung sowie Akzenten in patiniertem Silber sorgt die mit Leder besetzte Instrumententafel für eine exklusive Atmosphäre. Hervorzuheben ist in diesem Kontext auch eine neu konzipierte User Interface, die das Center Instrument als zukunftsweisende Möglichkeit der Interaktion bietet.
/
MINI presented the Clubman Concept – an elegant car with a length of 4.22 m and a width of 1.84 m in warm berry red and 'Blade' (a metallic grey colour) – at the 2014 Geneva motor show. Its dimensions exceeded the current MINI Clubman. While the first 1959 Mini was mainly characterized by maximum space in minimum dimensions, this ambition was now transferred to a new – and larger – category of car.

The Clubman Concept marked MINI's entry into a sector defined primarily by functionality and usefulness. An extraordinary design, distinctive material and colour mixes, and numerous clever ideas combine with the typical MINI exterior design features. Most notable is the elongated roof line indicating the car's large boot space, four doors, generous lateral surface gradients and a broad tail. The typical rear split doors were also retained, while accentuated here by the all-round window design. Combined with the horizontal tail lights, this creates a modern overall appearance. The MINI Clubman Concept also includes numerous aerodynamic features, such

as roof spoiler, the AirBreather and AirCurtains, with all openings and outlets being integrated into the car body.

The interior is characterized by extremely high-quality materials. As a key design element the ellipse is the most striking. Together with the sumptuous wood and leather interior with patinised silver accents, the leather finish of the dashboard create an exclusive atmosphere. The new user interface that turns the centre instrument into a forward-looking interaction space is another key feature.

Was bisweilen noch als Zukunftsmusik erscheint – für MINI-Fans ist MINI Connected bereits Realität. Sobald der Nutzer sein Smartphone über die USB-Schnittstelle mit dem Wagen verbindet, versorgen die MINI Connected Apps die Passagiere mit einem breiten Angebot unterschiedlichster Funktionen. So findet man mit der Online-Suche der MINI Connected APP umgehend ein Restaurant nach seinem Geschmack. Hat das Display die verfügbaren Ziele angezeigt, genügt ein Knopfdruck und die betreffende Adresse wird umgehend dem MINI-Navigationssystem übermittelt. Genauso funktioniert es mit dem Autoradio. Spielt es nicht die Lieblingsmusik, findet MINI Connected immer einen Sender, der gefällt. Dank Webradio

ist man unabhängig von UKW-Frequenzen und kann Tausende von Radiostationen aus dem Internet empfangen.

Besonders viele Vorteile bietet auch der MINI Connected XL Journey Mate, ein inzwischen oft unverzichtbarer Beifahrer, der den Fahrer bei der Planung einer Reise unterstützt, ihm verschiedene individuell zugeschnittene und kontext-bezogene Informationen liefert, ans Tanken und an geplante Telefonate erinnert oder bei der Suche nach einem Parkplatz hilft.

Was in Science-Fiction-Serien gern geboten wird, ist für MINI Connected bereits Wirklichkeit: Die Funktion Mission Control verwandelt das Fahrzeug in eine sprechende Persönlichkeit und stellt eine intensive Interaktion mit

dem Fahrer her. Will dieser besonders spritsparend unterwegs sein, steht er in direktem Austausch mit dem Bordcomputer und nimmt nützliche Hinweise von einer Stimme aus dem Off entgegen.

Einen besonderen Service bietet auch der Driving Excitement Analyser. Dank dieser App-Funktion sammelt der Fahrer Wertungspunkte beim Beschleunigen, Schalten, Lenken und Bremsen, er erfährt, wie sicher und gekonnt er das Potenzial seines Wagens nutzt, und wird für sportliches und zugleich sicheres Fahren, präzise Gangwechsel, kontrollierte Bremsvorgänge und harmonisches Kurvenfahren honoriert. MINI Connected macht den eigenen MINI im neuen Zeitalter kollektiver Vernetzung zum individuellen

Zentrum von Information und Kommunikation – nicht nur, was den Kontaktaustausch via Facebook und Twitter anbelangt.

/

It may sound futuristic but for MINI fans MINI Connected is already part of the real world. When a user connects their smartphone to the car via the USB interface, MINI Connected apps provide passengers with an extensive variety of functions. For example the MINI Connected app can quickly search online to find a suitable restaurant. Once the display shows what is available, the chosen address is transmitted immediately to the MINI navigation system at the touch of a button. The same applies for the car radio. If a radio channel is not playing your

favourite music, MINI Connected will find a radio station that does. Web radio does not rely on FM frequencies and can find thousands of radio stations on the Internet.

The MINI Connected XL Journey Mate also offers countless benefits and has developed to become an indispensable companion, helping drivers to plan their journey by delivering customized context-related information, reminding the driver to buy petrol or make scheduled phone calls or assisting in locating a parking space.

MINI Connected has made science fiction a reality. The Mission Control function turns the car into a talking personality that can interact with the driver. If the driver wants to save fuel while driving, they

can communicate directly with the on-board computer, whose voice responds with helpful hints.

The Driving Excitement Analyser also offers a special service. This app function enables the driver to collect points for accelerating, gearshifting, steering and braking, and thus they can find out how safe and skilled they are in using the car's potential. The driver is rewarded for a safe and sporting drive, precise gear changes, controlled braking and smooth cornering. MINI Connected turns the MINI into an individual information and communication centre for the new age of collective networking – and not just for checking your Facebook and Twitter accounts.

Wer hätte je gedacht, dass ein kompakter Kleinwagen zu Beginn des 21. Jahrhunderts im hart umkämpften Automobilmarkt derart einschlägt. Viele Kritiker hatten der Marke mit dem »Besitzerwechsel« 1994 und der Markteinführung des neuen MINI im Jahr 2001 das sichere Aus prophezeit. Inzwischen hat sich der MINI etabliert: Er ist ein cleveres Raumwunder wie einst, ein guter Freund und Ausdruck eines modernen Lebensgefühls. Heute kann man der Marke bescheinigen, dass sie an Unkonventionalität und Eigensinn nicht verloren hat. Vielfältig wie in den Sechzigerjahren ist ihre Modellpalette. Milieuforscher haben frühzeitig erkannt, dass sich ein bestimmter Personenkreis mit dem MINI neuer Prägung am meisten identifiziert: Es sind die Kreativen und Designaffinen, die als »selbstbewusste Lifestyle-Architekten« ihren Lebensstil selbst schneidern und als Lebensmittelpunkt die Metropole wählen. Sie provozieren Widersprüche, probieren gern Neues aus, kombinieren und überraschen. Ihnen gemeinsam ist der Wunsch, sich zu differenzieren und ihrer Individualität Ausdruck zu verleihen. Damit gelten sie als Trendsetter eines modernen Lebensstils. Der MINI greift dieses Lebensgefühl auf. Auch er ist voller Widersprüche: einerseits verhältnismäßig klein und kompakt, andererseits ein Raumwunder, einerseits zum begehrenswerten Premiumprodukt gereift, andererseits sympathisch und vertraut »like a friend«, mal ikonenbehaftet und traditionsbewusst, mal ultramodern, mal sportlich aktiv und extrovertiert, mal ganz durchdrungen von britischem Understatement.

Es ist zu kurz gegriffen, den MINI des 21. Jahrhunderts als reines Lifestyle-Produkt zu kategorisieren. Vielmehr ist er die Inkarnation eines selbstbewussten, die Gegenwart genießenden Lebensstils. Sein Design, sein Fahrvermögen, seine Kompaktheit macht ihn attraktiv und einzigartig. Die, die ihn heute entwerfen und stets neu erfinden, haben nicht nur den wirtschaftlichen Erfolg im Blick. So verspricht der derzeitige MINI-Geschäftsführer Jochen Goller: »MINI setzt sich zum Ziel, die erfinderischste und vom Kunden besessenste Automobilmarke der Welt zu sein. MINI steht seit 1959 für clevere Raumausnutzung, Trends zu setzen und ein Sportler durch und durch zu sein.«

MINI HEUTE
/
MINI TODAY

Who would have thought that a compact small car could be such a massive success in the highly competitive automotive market at the beginning of the 21st century? Quite a few critics had forecasted the end of the brand when the owner changed in 1994 and the new MINI was introduced in 2001. Since then the MINI has established itself. It is still as much a miracle of space as before. It is a good friend and an expression of a modern way of life. Today the brand proves that it has lost nothing of its unconventionality and tenacity. The model range is as varied as it was back in the 1960s. Researchers discovered early on that a certain group of people identifies most with the new MINI: they are creative, design-oriented 'self-confident lifestyle architects', who invent their own lifestyle and choose the big city as their home. They revel in contradictions, like trying out new things, adore combinations and surprises. These people share a common desire to be different and to express their individuality. This makes them trendsetters of a new modern lifestyle. The MINI embraces this way of life. The MINI is likewise full of contradictions: relatively small and compact on the one hand and a miracle of space on the other hand; established as a desired premium product but at the same time sympathetic and familiar 'like a friend'; sometimes iconic and traditional and sometimes ultramodern; now sporty and extrovert, now imbued with British understatement.

The 21st-century MINI is more than just a mere lifestyle product, however. It is much more the incarnation of a self-confident lifestyle that embraces the present. Its design, its driving performance and its compactness make the MINI attractive and unique. Those who design and constantly re-invent the MINI today do not focus only on commercial success. Jochen Goller, VP of MINI Sales promises: 'MINI aims to be the world's most inventive and customer-obsessed automotive brand. Since 1959 MINI has stood for the clever use of space, setting trends, and being a sports star through and through'.

GALERIE
/
GALLERY

Morris Mini-Minor
1959

Morris Mini-Minor
1959

Mini Ice Cream Van

Baujahr unbekannt / Construction date unknown

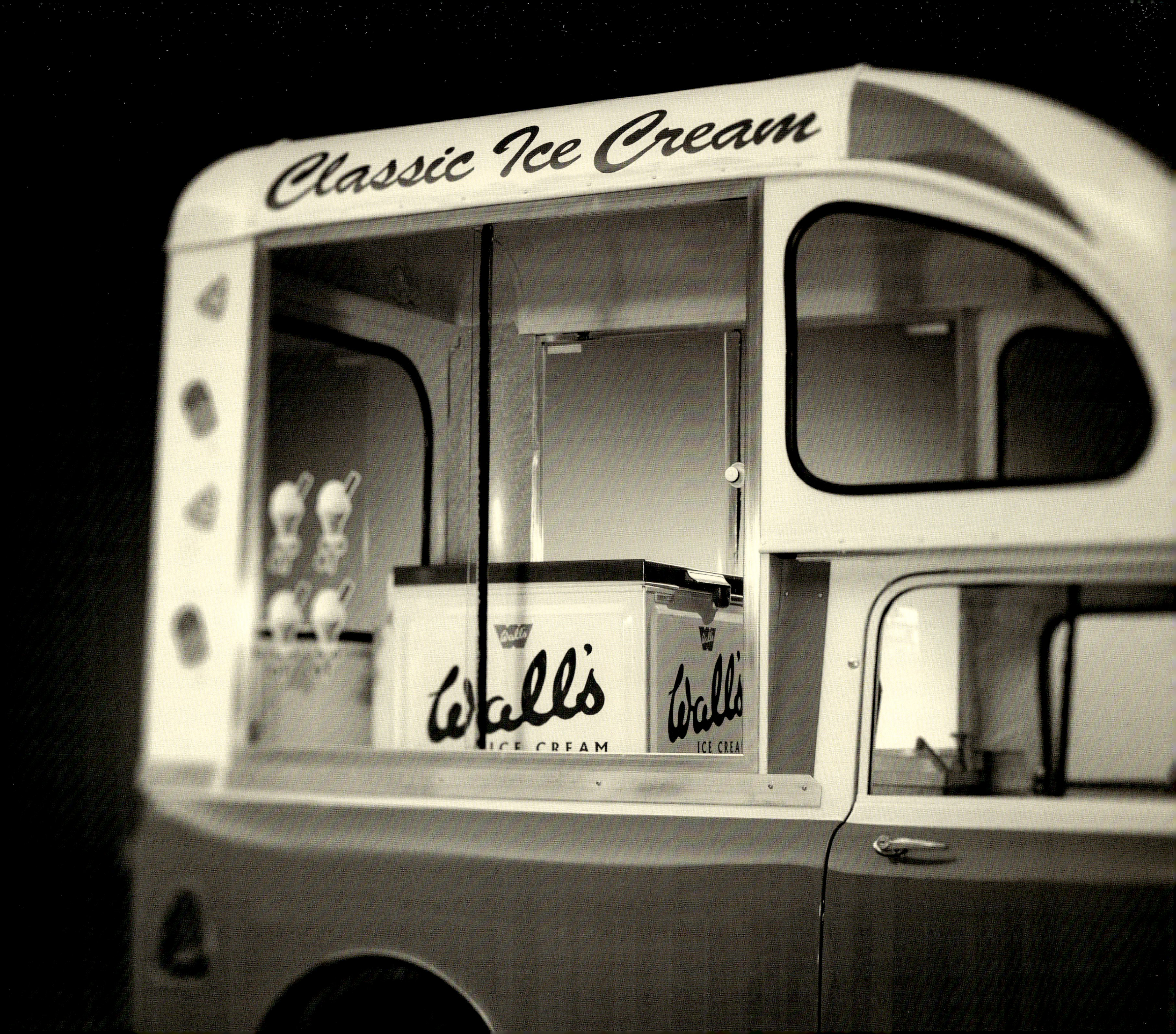

Classic Ice Cream
Wall's
ICE CREAM
Wall's
ICE CREA

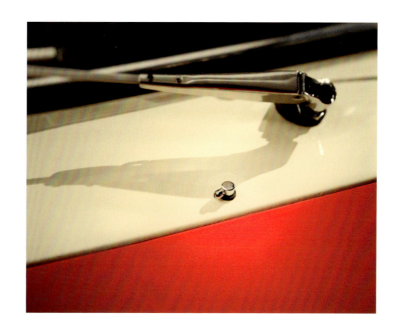

Mini Ice Cream Van

Baujahr unbekannt / Construction date unknown

Morris Mini Traveller

1960

Morris Mini Traveller

1960

Morris Mini Traveller

1960

Riley Elf
1969

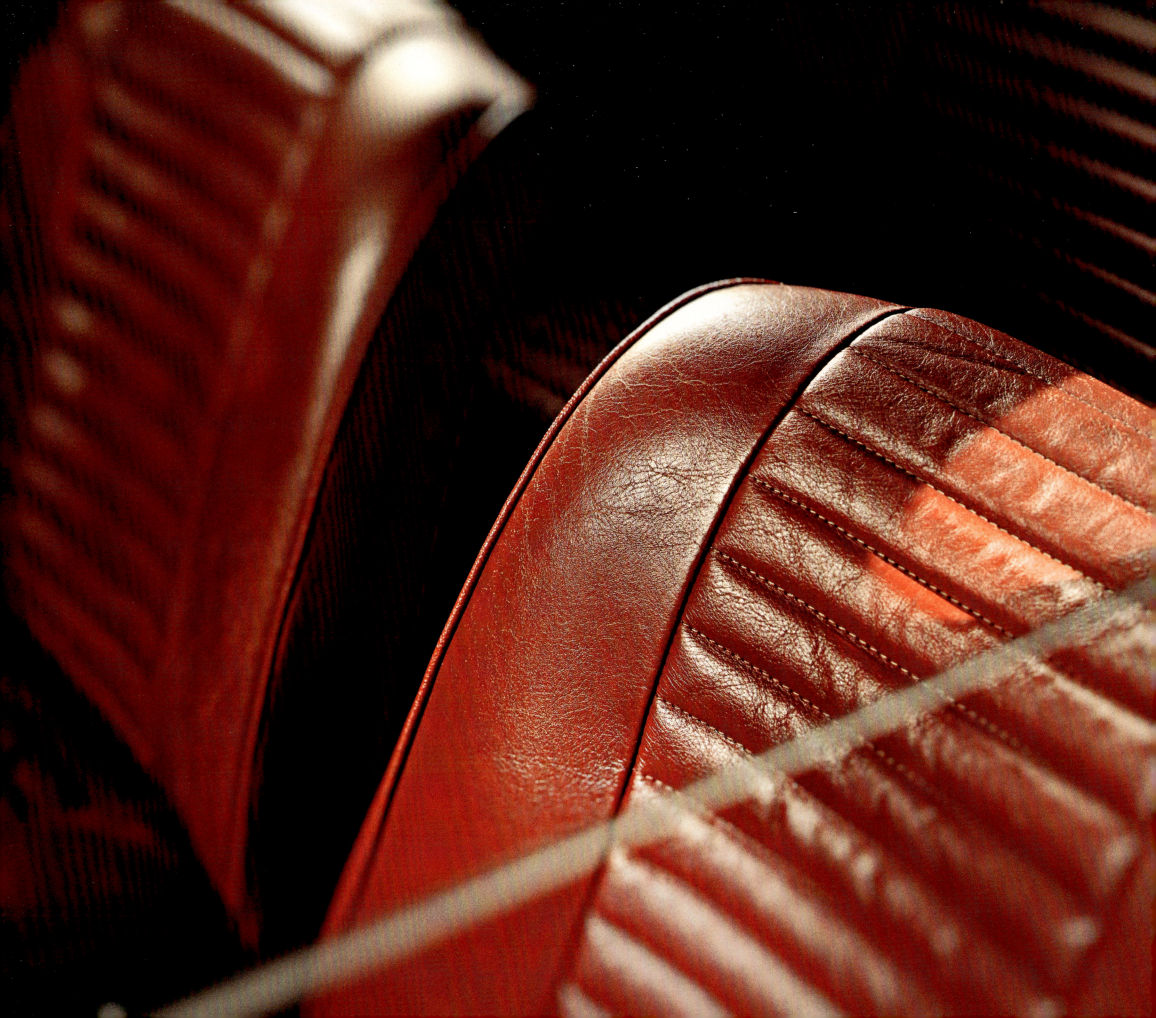

Mini Shorty

1963 / 2002

Mini Shorty

1963 / 2002

Mini Shorty

1963 / 2002

Mini Moke
1964

Mini Moke

1964

Mini Cooper S

Monte Carlo Siegerfahrzeug / The winning car at Monte Carlo

1967

Mini Clubman

1969

Mini Clubman

1969

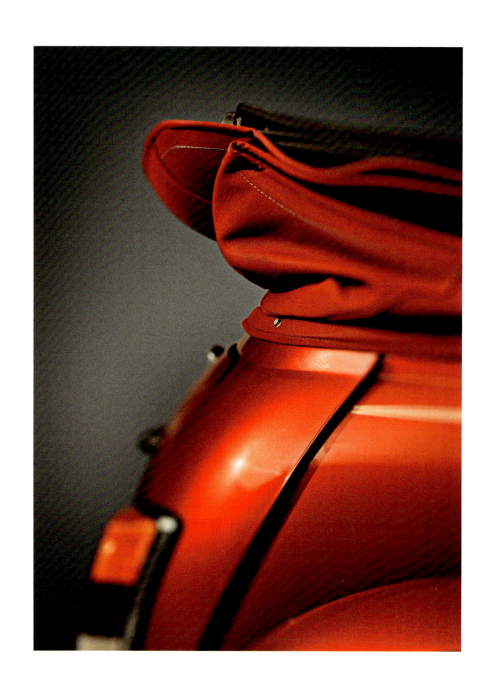

Mini Cabrio / Convertible

1991

Mini Cabrio / Convertible

1991

MINI ACV 30

1997

MINI ACV 30

1997

The New MINI

2001

The New MINI

2001

MINI Cabrio / Convertible

2004

MINI Cabrio / Convertible

2004

MINI Cabrio / Convertible

2004

MINI Cooper S Challenge »Alessandro Zanardi«

2005

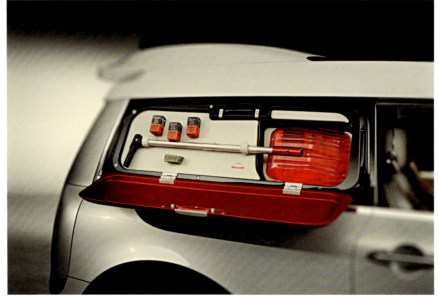

MINI Geneva Concept

2006

MINI Geneva Concept
2006

MINI Clubvan BBQ

2007

MINI Clubvan BBQ

2007

MINI Clubvan BBQ

2007

MINI Coupé Concept
2009

MINI Coupé Concept

2009

MINI Countryman

2010

MINI Countryman

2010

MINI Roadster

2011

MINI Roadster

2011

MINI Rocketman Concept

2011

MINI Rocketman Concept

2011

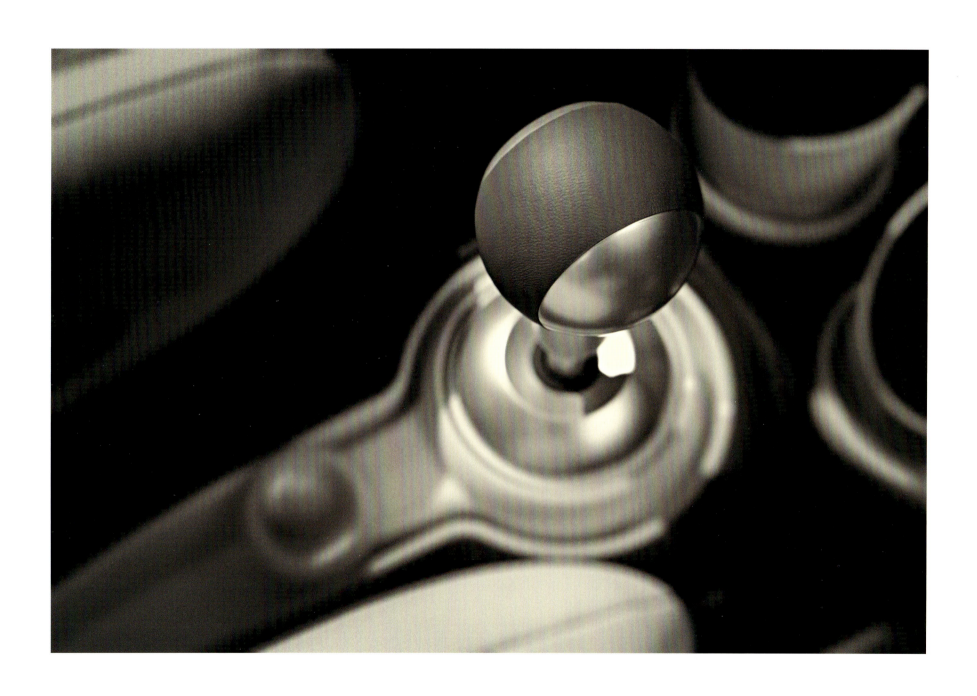

MINI Rocketman Concept

2011

MINI Paceman

2013

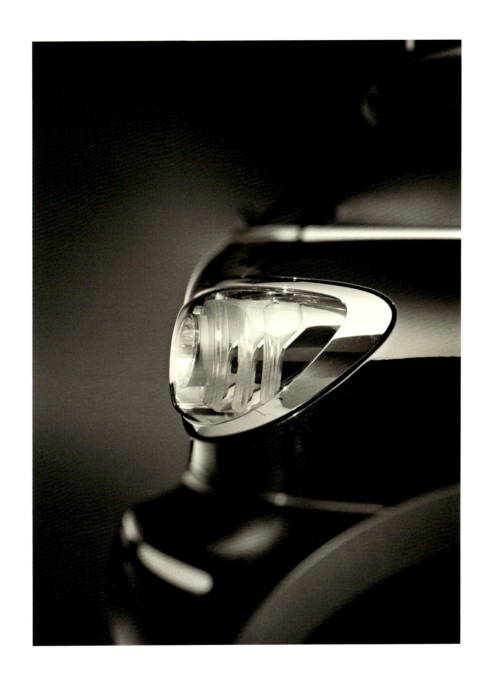

MINI Paceman

2013

Aufnahmen von | Photographs by
Erik Chmil

Der Herausgeber dankt allen Leihgebern, die zum Gelingen der Ausstellung und des Katalogs beigetragen haben / The Editor is grateful to all the lenders who have contributed to the success of the exhibition and this catalogue

Stephan Kirsten, Kevin Palmer, Trevor Ripley, Dominik Schaidnagel, Patrick Stellwag

Timorous Beasties, Glasgow
Ingo Maurer, München / Munich
Magis Design, Torre di Mosto
Die Neue Sammlung – The International Design Museum, München / Munich
TIM Textil- und Industriemuseum / Augsburg
Victoria & Albert Museum, London
Vitra Design Museum, Weil am Rhein

Design Icons S. / p. 31

Leihgaben / Loans:
Die Neue Sammlung – The International Design Museum Munich 1 Nick Roericht, Hotelstapel-geschirr / stackable hotel tableware »TC 100«, Thomas / Rosenthal, 1958 / 59. 2 Joe Colombo, Tischleuchte / table lamp »Spider«, O-Luce, 1966. 4 Aldo Van den Nieuwelaar, Tischleuchte / table lamp »TC-2 Radius«, Nila Lights, 1969. 7 Helmut Bätzner, Stapelstuhl / stackable chair »Bofinger-Stuhl BA 1171«, Menzolit-Werke, 1964. 8 Yves Behar, Lautsprecher / loudspeaker »Jambox«, Jawbone, 2010. 10 Marco Zanuso / Richard Sapper, Tragbares Transistorradio / portable transistor radio »Ts 502«, Brionvega, 1964. 12 Apple Industrial Design Team (Jonathan Ive), iPod MP3-Player, Apple Computer, 2001. 13 Verner Panton, Stapelbarer Kunststoffstuhl / stackable plastic chair »Panton Chair«, Herman Miller Furniture Company, 1960–1967, Ausführung / model 1973. 14 Hans Gugelot, unter Mitwirkung von / in collaboration with Dieter Rams, Radio-Plattenspielerkombination / radio / record player »Phono-Super SK4«, Braun, 1956. 15 Perry A. King / Ettore Sottsass, Reiseschreibmaschine / portable typewriter »Valentine«, 1969. 16 Konstantin Grcic, Service / tableware »Coup«, Thomas Rosenthal Group, 2003.
Magis Design 3 Konstantin Grcic, »Chair One«, 2003.
Ingo Maurer 5 Ingo Maurer, Tischleuchte / table lamp »Pollux«, 1967. 6 Ingo Maurer, Tischleuchte / table lamp »I Ricci Poveri – Toto«, 2014. 9 Moritz

Waldemeyer, Ingo Maurer und / and Team, Stehleuchte / floor lamp »My New Flame«, 2013. 11 Ingo Maurer, Tischleuchte / table lamp »Bulb«, 1966

Bildnachweis / Picture Credits

© British Motor Industry Heritage Trust, Gaydon Abb. S. / figs. p.: 15, 16, 18/19, 20, 23, 37, 49, 50, 51, 55, 56, 58, 126, 128, 131. © BMW AG, München / Munich Abb. S. / figs. p.: 2, 8, 13, 17, 21, 26, 28, 29 links / left (Foto / photo: Walter Bayer), 30, 44, 45, 47, 60, 72, 74, 75–77, 80, 85, 86, 89, 93, 95–97, 99, 101–103, 108, 109, 110–124, 127, 129, 133, 135, 139, S. 140–148, 148 /Mini goes to Tibet (Foto 2 / photo 2: Car Owner Magazine, Weih Zhang, Foto 3 and 4 / photo 3 and 4: Top Gear China Magazine, Chaucer K.C.Wong), 149 /Mini goes to Tibet (Foto 1 / photo 1: China Auto Pictorial Magazine, Shang Liu, Foto 2 / photo 2: Car Owner Magazine, Wei Zhang), 150, 151, Umschlagrückseite / back cover; S. / p. 156–233 und / and Umschlag-vorderseite / front cover (Fotos / photos: Erik Chmil). © BMW Group Archiv Abb. S. / figs. p.: 41, 52–54, 57, 59, 61, 65, 66, 67, 68–71, 79, 82, 83, 87, 88, 91, 105–107, 132, 136–138. © Getty Images Abb. S. / figs. p.: 27 (Foto / photo: Keystone Features), 33 Mitte / centre (Foto / photo: Bentley Archive / Popperfoto), 34 (Foto / photo: Paul Popper / Popperfoto / Getty Images), 35 (Foto / photo: Roger Jackson), 36 (Foto / photo: GAB Archive), 39 (Foto / photo: Popperfoto). © Hulton-Deutsch Collection / CORBIS Abb. S. / fig. p.: 33 oben und unten / top and bottom. © Die Neue Sammlung – The International Design Museum, München / Munich Abb. S. / figs. p.: 131 (Fotos / photos: A. Laurenzo). © Paramount Film Studios Abb. S. / figs. p.: 24 oben rechts / top right, S. 24 Mitte links / left centre, unten links / bottom left. © Picture-Alliance / Photoshot Abb. S. / fig. p.: 68. © Magis Abb. S. / fig. p.: 131. © Trevor Ripley Abb. S. / fig. p.: 38. © Magis Abb. S. / fig. p.: 131. © Ingo Maurer Abb. S. / figs. p.: 131. © VDL Ned-car Abb. S. / figs. p.: 62, 63. © Universal Film Studios Abb. S. / figs. p.: 24 Mitte rechts / right centre, unten rechts / top right. © Nick Veasey Abb. S. / fig. p.: 29 rechts / right. © Warner Brothers Distributing INC. Abb. S. / figs. p.: 24 oben links / top left

Künstlerrechte / Artist rights
© VG-Bildkunst Bonn, 2014 für das Werk von Ettore Sottsass
© Christian Claerebout
© Nick Veasey
© Vitra für den »Panton Chair«
sowie bei den jeweiligen Künstlern und ihren Rechtsnachfolgern

Der Herausgeber hat sich intensiv bemüht, alle Inhaber von Abbildungs- und Urheberrechten ausfindig zu machen. Personen und Institutionen, die möglicherweise nicht erreicht wurden und Rechte beanspruchen, werden gebeten, sich nachträglich mit dem Herausgeber in Verbindung zu setzen. / The editor has made every effort to identify all copyright holders. The publisher would be grateful to receive notification from any persons or institutions with copyright claims whom it may not have been possible to reach.

Zitat- und Quellennachweis / References

S. 34 / p. 33 Rainer Metzger, *Swinging London*, Wien 2011, S. 66 / Rainer Metzger, London in the Sixties, trans. David H. Wilson (London: Thames and Hudson, 2012), p. 65; Jürgen Kramer, »Der Mini und der Anfang vom Ende der britischen Autoindustrie«, in: Claus-Ulrich Viol (Hrsg. / Ed.), *Mini & Mini, Ikonen der Popkultur zwischen Dekonstruktion und Rekonstruktion*, Bielefeld 2009, S. / p. 29 / Englische Übersetzung von / English translation by Philippa Hurd
S. 38 / p. 38 Viola Hofmann, »Their own teenage look?«, in: Claus-Ulrich Viol (Hrsg. / Ed.), *Mini & Mini, Ikonen der Popkultur zwischen Dekonstruktion und Rekonstruktion*, Bielefeld 2009, S. / p. 42
S. 65 / pp. 64–65 Wolfgang Sachs, *Die Liebe zum Automobil*, Reinbek 1984, S. 80 / Englische Übersetzung von / English translation by Philippa Hurd

Impressum / Imprint

Diese Publikation erscheint begleitend zur Ausstellung / This book is published to coincide with the exhibition

»The MINI Story«
BMW Museum, München / Munich
27.11.2014–31.1.2016

Herausgegeben von / Edited by
Andreas Braun, BMW Museum, München / Munich

Projektassistenz / Project assistant
Nathalie Schwind

Projektmitarbeit / With the collaboration of
Theresa Czerny, Nicole Prinz, Anna Schleypen, Florian Moser, Hannes Ziesler

Fachlektorat / Technical Advisor:
John Keith Adams, Kettering, UK

Projektmanagement / Project coordination,
Hirmer Verlag: Jutta Allekotte, Peter Grassinger

Lektorat Deutsch / German copy-editing:
Büro Anne Funck, München / Munich

Übersetzung / Translation:
Telelingua und / and Philippa Hurd

Lektorat Englisch / English copy-editing:
Philippa Hurd, London

Produktion / Production: Peter Grassinger

Lithografie / Prepress and repro: Repromayer Medienproduktion, Reutlingen

Druck und Bindung / Printed and bound by:
Printer Trento S.r.l., Trento

Printed in Italy

Bibliografische Information der Deutschen Nationalbibliothek: Die Deutsche National-bibliothek verzeichnet diese Publikation in der Deutschen Nationalbibliografie; detaillierte bibliografische Daten sind im Internet über http://www.dnb.de abrufbar. / Bibliographic information published by the Deutsche National-bibliothek: The Deutsche Nationalbibliothek lists this publication in the Deutsche Nationalbiblio-grafie; detailed bibliographic data are available in the Internet at http://www.dnb.de.

© 2014 Hirmer Verlag GmbH, München / Munich und / and BMW AG, München / Munich

ISBN 978-3-7774-2402-6 (Deutsche Ausgabe)
ISBN 978-3-7774-2372-2 (English edition)

www.hirmerverlag.de
www.hirmerpublishers.com